SALESTEGY

Combining SALES & STRATEGY –
Guiding Innovative Business
Models to Success

Jan Wieseke

Translation: Ian Rhodes
Copy editor: Robert Furlong

Cover design: zweiband.media, Berlin
Layout, design, and illustrations: zweiband.media, Berlin
Back cover photograph: © Daniel Sadrowski

Printing and binding: Beltz Grafische Betriebe GmbH, Am Fliegerhorst 8, 99947 Bad Langensalza, Germany
Published by Bochum Sales Publishing GmbH, Erasmusstrasse 39, 42653 Solingen, Germany

ISBN: 978-3-9823858-3-9

For my family – Christina, Emil & Mattis

Thanks

When you read these thanks and compare them with those in the first volume about the Sales Profit Chain,* you will discover that many of the passages are identical. This is because the two books are very closely related. I was extremely fortunate to be able to count on terrific support for both volumes. When I repeat passages that I already wrote in Volume 1, I do so with a deep sense of gratitude.

When my friend and colleague Prof. Olaf Plötner advised me to finally start writing "YOUR sales management book" and I then decided to set to work, I had no idea what I was letting myself in for. I also had no idea that it would ultimately turn out to be two volumes. During the seven years it took to write, I reached the limits of my personal endurance several times. However, working on both volumes has given me an even greater passion for sales management.

Writing the book took up every spare minute of my time. In the final analysis, these added up to thousands of hours, but that alone would never have sufficed to complete the two volumes. I needed active support.

The two volumes would never have seen the light of day without the 30-person team of our Sales Management Department at Ruhr University of Bochum. Nobody can get to grips with 20 years of sales research involving more than 4,000 companies on their own. Our team in Bochum is driven by a love of the subject. I would like to thank them all from the bottom of my heart. When I use the term "we" in the two volumes, it always refers to this team. I would particularly like to thank Dr. Marvin Brundtke, Prof. Johannes Habel, Dr. Bianca Richter, Dr. Hanaa Ryari, and Prof. Sabrina Scheidler for their all-round support during every phase of the writing process. I would like to thank Jan Altenscheidt, Dr. Jacqueline Baudach, Dr. Maximilian Dax, Dr. Christina Desernot, Dr. Maximilian Frieß, Dr. Lukas Isenberg, Dr. Verena Maag, Jonas Rübertus, and Dr. Dennis Schendzielarz for their tireless work on individual modules. For their support during the correction loops, I would like to thank Maximilian Bäker, Sophie Carneim, Felix Dauerer, Jule Delventhal, Philipp Giordani, Leah Kurbjuhn, Hanna-Lena Paulus, and Teresa Pietrzinski. A big thank you also goes to our secretaries, Mira Buchholz and Caterina Ledwig, for their support in everything that needed doing for the two volumes.

I would also like to thank my friends Dr. Tino Kessler-Thönes and Dr. Thomas Lenz. You stood by me with advice for, and support with, many detailed questions whenever the need arose.

* Wieseke (2022, Volume 1): *The Sales Profit Chain – Understanding Causal Chains, Optimizing Sales, Increasing Profitability*, Bochum Sales Publishing.

This work would not have come about without my colleagues in our department – Prof. Sascha Alavi and Prof. Christian Schmitz. I am grateful that you are not only colleagues, but also genuine friends. Without your support, I would not have been able to establish the SALESTEGY logic.

Unfortunately, my thanks come too late for my deceased colleague and friend, Prof. Mario Rese. His ideas were far ahead of their time, and they inspired many of the thoughts that I have incorporated into this work.

Without the practical examples lifted from the project experience of our consultancy service *Prof. Schmitz & Wieseke – Sales Management Consulting*, both volumes would have been lacking in substance. It is this rich experience and the numerous practical examples that make the concepts tangible in the first place. My thanks particularly go to Dr. Helge Guba and Lukas Haxter.

Without professional translation and editing as well as graphic design, the two volumes would not look the way they do. They would have been full of mistakes and awkward to read. I had the best support anyone could wish for. Many thanks to Ian Rhodes for his smooth translation and to Robert Furlong for his precise editing. Thanks also to Daniel Bauer and Doreen Wilhelm of *zweiband.media* for the graphic design and typesetting. This was fantastic.

Nor would the two volumes have come about without our numerous corporate partners. I would like to thank all of the companies who have sharpened the SALESTEGY logic with us in workshops and provided us with their confidential sales data for detailed analyses. Hopefully, the occasionally painful feedback loops have made the result usable for numerous sales decision-makers. My particular thanks here go to Dr. Felix Brüggemann, Bastian Brunow, Holger Jentsch, Marc Kühn, and Sebastian Springer. I have Michael Sindram to thank for the most intensive practical sparring for the two volumes. I have said "many thanks" to him so often that, by my reckoning, he is already a thank-you millionaire. His feedback substantially influenced my work.

Thank you all from the bottom of my heart!

Contents – SALESTEGY (Combining SALES & STRATEGY)

PART 1 – WHY SALESTEGY?

PART 2 – STRATEGY – FINDING THE RIGHT BUSINESS MODELS

PART 3 – PRACTICAL EXAMPLES

PART 4 – SALES – IMPLEMENTING NEW BUSINESS MODELS IN SALES

Executive summary

SALESTEGY is the combination of SALES & STRATEGY.
It is becoming increasingly important due to the spread of Death Valley markets.

Why are Death Valley markets spreading?

Death Valley markets are market sectors in which suppliers are only able to sell their products and services with very low profit margins. Under such market conditions, the vital margin for companies is in equally short supply as water is for plants in Death Valley.

Many companies from highly developed industrialized nations are increasingly finding themselves in the maelstrom of Death Valley markets. This process is picking up pace as a result of globalization, digitalization, and crises.

How can companies reach green hill markets?

Green hill markets are market sectors in which customers are willing to pay for a company's products and services at a price that enables the supplier to achieve a healthy profit margin.

To achieve the objective of a high willingness to pay, companies have to develop intelligent solutions for their customers. Many B2B companies are therefore increasingly investing in **solution business models**. Only 43% of the B2B companies analyzed by us still regard themselves as being in a pure product business.

What does "SALESTEGY" mean?

SALESTEGY

SALESTEGY is the combination of SALES & STRATEGY. Solution business models require this combination, both to **find** them and to **implement** them.

SALESTEGY uses sales in a dual role: 1) As a catalyst during the finding process and 2) as an implementer in introducing solution business models.

What changes as a result of SALESTEGY?

Old:

New:

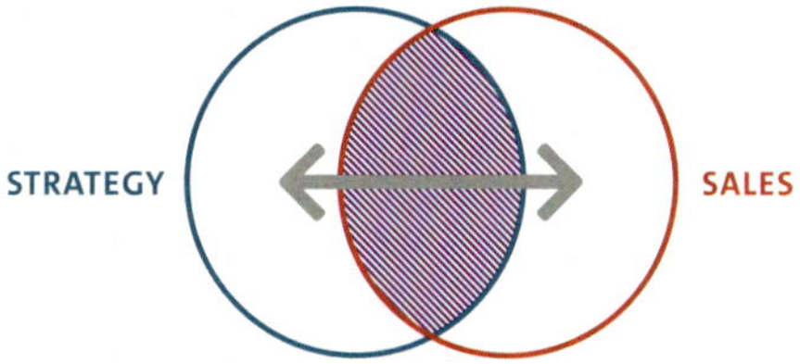

The concept of SALESTEGY runs counter to the classic distribution of roles, according to which strategic management and sales act remotely from one another, and the sales department acts according to the strategic management's specifications.

Conversely, the SALESTEGY role distribution provides for the close intermeshing of strategic management and sales. Here, sales acts on an equal footing, since sales representatives are involved in strategic decisions or are even permanently integrated at board level.

What does the STRATEGY area of SALESTEGY involve?

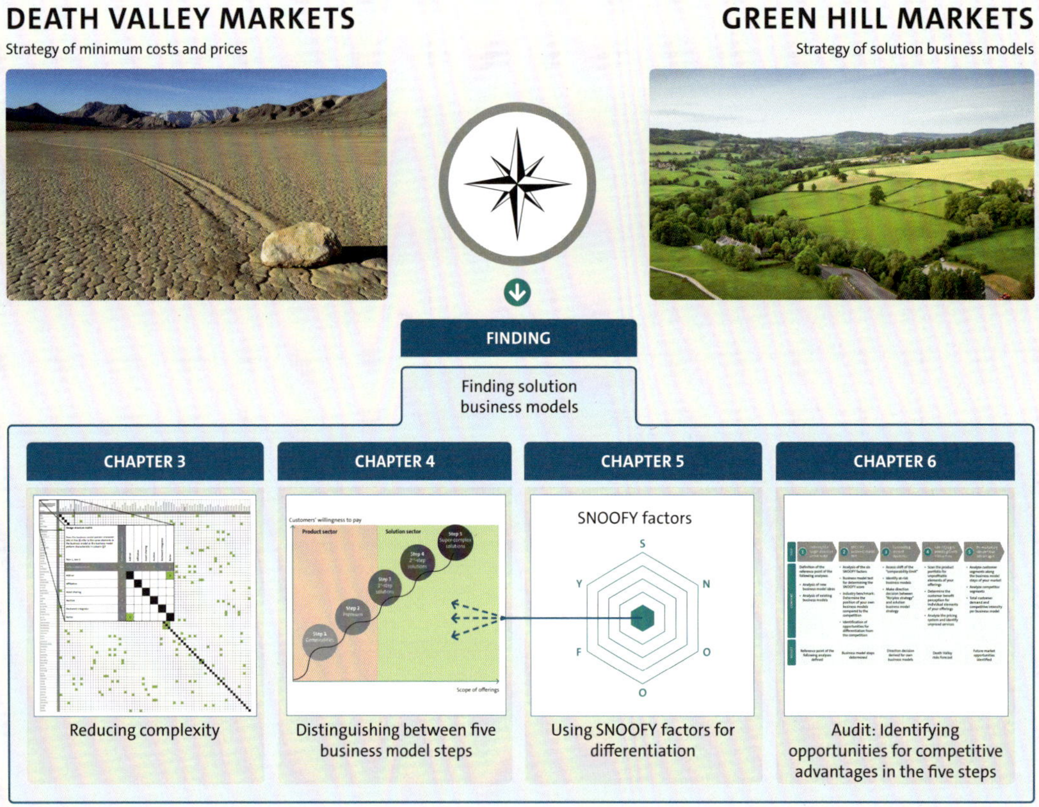

The STRATEGY area of SALESTEGY involves **finding** promising business models (see Chapters 3–6).

To do this, a logic that can be used to boil a variety of possible business models down to a manageable level is first required. The logic in this book enables a distinction to be made between five business model steps that can be encountered in highly developed markets today (commodities, premium, 1st-step solutions, 2nd-step solutions, super-complex solutions).

With the **"SNOOFY business model audit,"** I introduce a procedure that you can use to evaluate your competitive advantages within the five business model steps.

What does the SALES area of SALESTEGY involve?

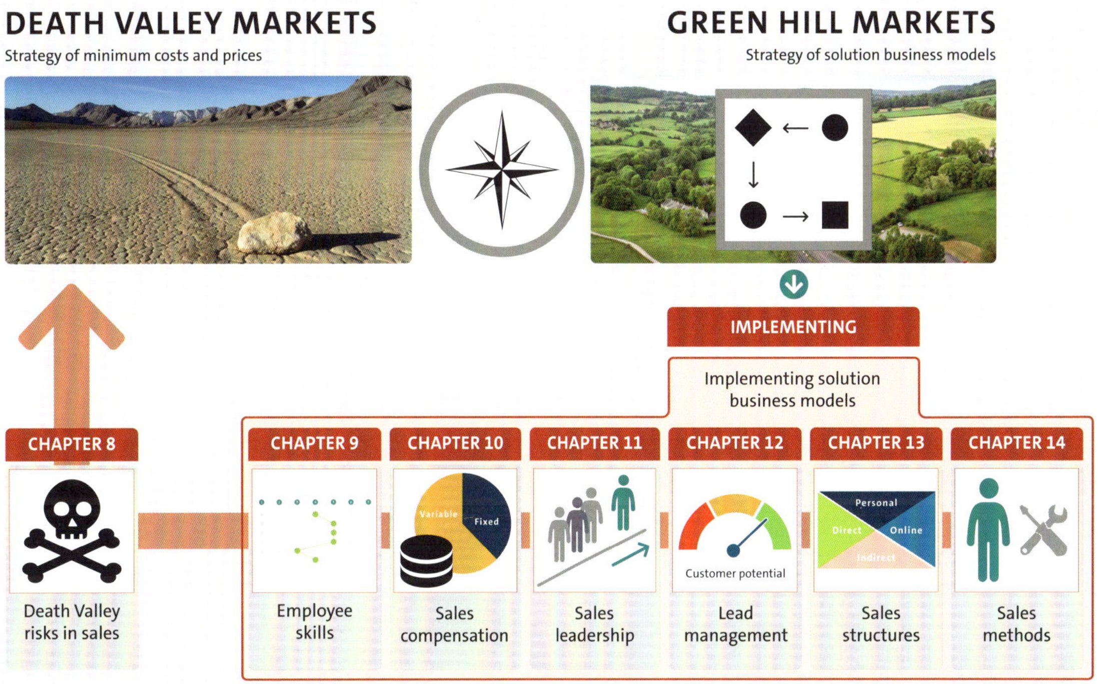

The SALES area of SALESTEGY involves **implementing** promising business models (see Chapters 9–14).

Entering a new business model step influences all areas of sales management, namely:

- which sales employee skills are necessary (see Chapter 9)
- which sales compensation is necessary (see Chapter 10)
- which sales leadership is necessary (see Chapter 11)
- which lead management is necessary (see Chapter 12)
- which sales structures (e.g., sales channels) are necessary (see Chapter 13)
- which sales methods are necessary (see Chapter 14)

However, there are "Death Valley risks" in sales (see Chapter 8). I argue that each of the above areas of sales management is so powerful that it can cause an entire business model to fail.

10 important terms in this book		
1	SALESTEGY	is the combination of SALES & STRATEGY. Solution business models necessitate the combination of the sales function and strategic management. This applies both to finding promising business models and implementing them.
2	Death Valley markets	are market sectors in which suppliers are only able to sell their products and services with very low profit margins. Under such market conditions, the vital margin for companies is in equally short supply as water is for plants in Death Valley.
3	Green hill markets	are market sectors in which customers are willing to pay for a company's products and services at a price that enables the supplier to achieve a healthy profit margin.
4	Commodity business models	are based on interchangeable standard products or services that are available in comparable quality from various suppliers (e.g., simple pipes and C-parts).
5	Premium business models	are founded on premium products or services based on quality aspects that the competition does not yet have (e.g., high-quality air conditioners).
6	1st-step solution business models	offer a benefit because suppliers make small process steps in the value chain easier for their customers (e.g., through after-sales services, training courses, maintenance contracts).
7	2nd-step solution business models	mean that suppliers intervene even more extensively in their customers' value chains. In this process, the suppliers actively assume responsibility for extensive process steps (e.g., warehouse management). Responsibility for the *results* of the processes is also assumed.
8	Super-complex solution business models	are the final step of the solution evolution. In these, suppliers extend their scope of offerings to such an extent that customers' core processes are shifted to them (e.g., software for managing diagnostic processes for patients in a hospital) and/or the supplier and customer are extensively integrated. The objective is maximum customer benefit.
9	Atriplex strategy	is a minimum-cost business model approach. All elements of a product or service that are not central to the target group are omitted (see examples on p. 145). This strategy mimics the minimalistic Atriplex desert plant in Death Valley.
10	SNOOFY factors	are six factors that can be used to distinguish between business models: ***S**ervitization, **N**urture customer processes, **O**perate processes for customers, **O**utcome-based pricing, **F**ar-reaching relationship, and **Y**ield customer value-in-use.*

WHY SALESTEGY?

STRATEGY – FINDING THE RIGHT BUSINESS MODELS

PRACTICAL EXAMPLES

SALES – IMPLEMENTING NEW BUSINESS MODELS IN SALES

Introduction

About the classification of this book

Message of this chapter

The objective of this book is to combine SALES & STRATEGY in order to guide innovative business models to success. To do this, the classic distribution of roles between strategic management and sales has to be broken down.

What should you have learned by the end of the chapter?

- In particular, the target group of this book is (sales) decision-makers in B2B companies.
- The scientific data basis of the book consists of 100 million qualified data points from 4,000 companies.
- It includes practical examples from 11 B2B industries (see industry directory, p. 23).

1 My appeal to you as the reader

This book is focused on the concept of SALESTEGY, or the combination of strategic management (STRATEGY) and sales management (SALES). It is aimed at clearly pointing out the importance of sales, both in establishing and implementing new, strategic business models. It is intended to show, step-by-step, which questions we have to ask ourselves in managing this process and where the challenges and risks lie.

This interface has often been neglected so far. Within companies, strategic management often acts divorced from sales and vice versa. In addition, the functions are frequently prone to silo thinking, which is directed against other functions, such as marketing vs. sales or product development vs. sales.

This silo thinking is also reflected in the literature. There are innumerable books about both strategic management and sales management (see Figure 1). To date, books that combine both of these areas are a rarity.

My appeal is: Let's overcome this silo thinking. Let's combine SALES & STRATEGY as SALESTEGY!

Fig. 1 The gap between the literature on SALES & STRATEGY

2 The target groups: For whom is this book helpful?

The target groups of this book are comparable to those of the companion book about the "Sales Profit Chain."[1] This book, too, is aimed at decision-makers who manage companies in dynamic B2B markets. Since it combines strategic management and sales management, it is aimed in equal measure at decision-makers from both areas. It shows the business models with which sustainable success can be achieved and the levers that have to be employed in strategic management and sales to do so.

The target groups that I want to reach with the SALESTEGY approach include all managers who are searching for new business models and have to implement these in the sales department. The first group consists of (sales) executives, managing directors, sales managers, and sales excellence coordinators. The second group encompasses future (sales) decision-makers, such as young professionals, sales employees, trainees, and students.

As the third group, external stakeholders can use this book to obtain an overview of the strengths and weaknesses of business models and their implementation in sales. In particular, this is of interest when an external company is to be acquired or capital is to be invested in a startup.

This target group focus clearly shows why I did not formulate this book in the form to which scientists are accustomed. I have attempted to avoid excessively complex descriptions. This may lead to a loss of scientific depth in some places. I therefore ask for the forgiveness of fellow researchers searching for details of the theoretical and analytical derivations. If I were to attempt to include these, the scope of this book would probably increase tenfold.

Nevertheless, our department's scientific investigations form the basis of this book. I would therefore like to address them briefly in the following.

[1] Wieseke (2022), *The Sales Profit Chain: Understanding Causal Chains – Optimizing Sales – Increasing Profitability*, Bochum Sales Publishing.

Target group	Benefits of the SALESTEGY approach
(Sales) decision-makers • (Sales) executives • Managing directors • Sales managers • Key account managers • Sales excellence coordinators	**Finding the *right* business models** • Conducting a business model audit to check where the greatest competitive advantage can be achieved in the future • Checking the steps of your business models and how they stand compared to the industry benchmark • Forecasting the dynamics of your market so as not to be caught out by disruptive developments **Implementing the right business models *correctly* in sales** • Adapting sales employee skills • Adapting sales compensation • Adapting sales leadership • Adapting lead management • Adapting sales structures • Adapting sales methods
Future (sales) decision-makers • Young professionals • Sales employees • Trainees, students	**Obtaining a basic understanding** • Understanding the mechanisms of innovative solution business models • Understanding the fundamentals of sales management in solution business models
External stakeholders • Banks • Acquirers • Consultants, M&A advisors • Startup investors	**Validating company valuations** • Checking the future promise of the business models of takeover candidates • Making sounder decisions concerning investments in companies/startups • Basis for SWOT analyses and due diligence

3 The scientific data basis of this book

As in the companion book about the "Sales Profit Chain," I used the findings that we have collected at the Sales Management Department at Ruhr University of Bochum during the past 20 years of our research and our cooperation with companies in writing this book. The scientific foundation is therefore an accordingly extensive data basis consisting of more than 100 million qualified data points from more than 4,000 companies.

The investigation that I would like to single out here is our business model study involving 845 companies. In this study, we analyzed the structures and success factors of business models in detail. The industry structure of the analyzed companies is shown in the adjacent table.

Industry distribution of the business model study involving 845 companies

Industry	Percentage
Industrial goods & mechanical engineering	26.6%
Pharmaceuticals & healthcare	5.7%
Energy & supply	2.5%
Construction	7.1%
Chemicals & base materials	2.6%
Technology & software	10.3%
B2B commerce & distribution	8.7%
Consumer & retail	5.8%
Financial services	5.7%
Automotive	5.7%
Transport & logistics	2.5%
Telecommunications	5.8%
Others	11.0%

4 The industry directory for the company examples in this book

Throughout this book, I draw on examples from various B2B industries. The following table provides an overview to help you look for the industry of interest to you.

Industry examples in this book

Industry	Pages
Industrial goods & mechanical engineering	36, 62, 64, 77, 83, 97, 107, 108, 109, 112, 113, 116, 117, 145, 148, 149, 155, 160, 169–171, 184–185, 188–193, 196–197, 204–205, 224, 243, 248, 249, 251, 252, 259–263, 278, 286, 288
Pharmaceuticals & healthcare	36, 64, 112, 152, 224, 248, 249, 313, 340
Energy & supply	36, 64, 115, 169, 176–177, 180–183, 186–187, 224, 248, 249, 317–321
Construction	36, 64, 112, 118, 120, 121, 160, 169, 192–193, 224, 248, 249, 338
Chemicals & base materials	36, 64, 112, 115, 145, 160, 224, 248, 249, 269
Technology & software	36, 64, 108, 109, 137, 152, 169, 172–175, 198–203, 224, 248, 249, 291, 293, 313, 338
B2B commerce & distribution	36, 39, 64, 148, 224, 248, 249
Financial services	36, 64, 79, 143, 224, 248, 249
Automotive	36, 64, 108, 115, 122, 160, 169, 178–179, 224, 247, 248, 249, 253, 286
Transport & logistics	36, 64, 115, 147, 169, 194–195, 224, 248, 249
Telecommunications	36, 64, 224, 234, 235, 248, 249, 297, 299

5 The basic structure of this book

The following figure shows the four parts of this book. They are arranged clockwise. Since the focal point is the concept of SALESTEGY, I determine why the combination of SALES & STRATEGY is becoming increasingly important for companies in the first part. The second part deals with the STRATEGY area of SALESTEGY. This is concerned with finding new solution business models that enable a sustainable competitive advantage. In the third part, I describe 18 practical examples of innovative solution business models. The fourth part deals with the SALES area of SALESTEGY, that is, with successfully implementing solution business models in sales.

Fig. 2 The four parts of this book

The aim of the first part of this book is firstly to determine why we have to deal with the concept of **SALESTEGY**, or the combination of **SALES** & **STRATEGY**. It is due to the spread of "Death Valley markets," in which customers' willingness to pay is barely sufficient for companies from highly developed industrialized nations any longer. In Chapter 1, I discuss that this is the reason why companies are seeking to escape to "green hill markets," where sufficient willingness to pay exists. They are attempting to do so by means of solution business models. A symbiosis of **SALES** & **STRATEGY** is necessary to find and successfully implement these. I therefore introduce the concept of **SALESTEGY** in Chapter 2.

WHY SALESTEGY?

STRATEGY – FINDING THE RIGHT BUSINESS MODELS

PRACTICAL EXAMPLES

SALES – IMPLEMENTING NEW BUSINESS MODELS IN SALES

Chapter 1

Survival in the age of Death Valley markets

Message of this chapter

The nature of Death Valley can be used as an analogy for many B2B markets in which margins are becoming increasingly meager.

What should you have learned by the end of the chapter?

- Many markets are turning into Death Valley markets, in which customers' willingness to pay is very low.
- The biggest driving forces behind Death Valley markets are: 1) Globalization, 2) digitalization, and 3) crises such as the coronavirus pandemic. These drivers act in concert and lead to the erosion of the willingness to pay.
- Only companies with minimalistic cost structures succeed in surviving in Death Valley markets. The majority of companies from highly developed industrialized nations are unable to survive in such environments.
- Escaping to green hill markets with a higher willingness to pay can be accomplished through solution business models.
- Compared to the classic product business, the aim of solution business models is to maximize the benefit to customers, and thus increase their willingness to pay.

1.1 A mental experiment: Just imagine you had to go into Death Valley

I would like to begin with a portrayal of Death Valley. It is located southeast of the Sierra Nevada (USA) and, with its heat, is symbolic of many other deserts. As can be seen from its barren, parched ground, it is an extremely hostile environment. With an average of just two days' rain per year, it is one of the driest places on Earth. And one of the hottest. In 1913, the world temperature record of 134.1°F was measured here. Even the average daily temperatures surpass the 105°F mark during the summer months. In July 2017, Death Valley established a new heat record. The average temperature throughout that month was 107.4°F, with 95°F being recorded at night and 119.5°F during the day.

These conditions pose a major problem for plants, animals, and humans. This is already indicated by the story of how Death Valley got its name. According to this, two groups of pioneers wandered disoriented through Death Valley for a number of weeks in 1849. In their desperation, they did everything they could to survive. They turned the wood from their trekking wagons into firewood and ate the oxen used to pull them. With the last of their strength, they succeeded in saving themselves over Wingate Pass. According to the story, the valley's name arose when one of the women in the group turned around a final time and shouted *"Goodbye, Death Valley!"*

Fig. 3 The nature of Death Valley

I would now like to ask you to take part in a mental experiment in which you yourself enter Death Valley. This mental experiment involves two scenarios.

Scenario 1: Requirements for surviving in Death Valley in the long term

Just imagine you are facing a mission requiring you to remain permanently in Death Valley. How would you prepare for it?

Consider for a moment which fundamental requirements would be needed for you to survive the mission unscathed.

How would you prepare for this challenge? Managers' responses:

"I'd undertake several desert expeditions as a trial in order to learn how to adapt to the conditions."

"I'd look for someone with experience of the desert as a coach."

"I'd buy the most important tools to be able to construct a well."

"I'd train to find food and water even in the desert."

"I'd practice my resilience to heat by specifically exposing myself to high temperatures."

"I'd train myself to survive with extremely little water."

Requirements for surviving in Death Valley in the long term

The managers' responses show the essential requirements that would have to be met in order to survive in Death Valley in the long term. One manager summed them up as follows: **"Perfectly adapted minimalists with tools for constructing a well would be ideal."** This means that the following are required:

- **Good adaptation to the extreme conditions** prevailing in Death Valley in terms of water, food, and temperatures.
- **Good equipment,** particularly for constructing a well and for protection against heat.
- **A good information basis and training** so that this adaptation is successful and the equipment is used optimally.

Fig. 4 Requirements for surviving in Death Valley in the long term

Scenario 2: Requirements for crossing Death Valley

Just imagine you are facing a mission requiring you to cross Death Valley on foot and without any electronic aids. How would you prepare for it?

Consider for a moment which fundamental requirements would be needed for you to survive the mission unscathed.

How would you prepare for this challenge? Managers' responses:

"I'd think about my equipment and make sure I had the right water backpacks, the right shoes, the right clothing, the right sun protection."

"I'd try to clearly determine the challenge: How long is the route? What are the temperatures? How much rainfall is there?"

"I'd check out various routes to find the one that's easiest to manage."

"I'd talk to people with desert experience who could give me instructions on how to proceed and survival tips such as orientation at night."

"Orientation would probably be crucial. So I'd buy myself a good, classic compass."

"I'd try to train my stamina as well as possible."

"I'd ask experts whether it would be better to walk during the day or at night."

Requirements for crossing Death Valley

The above responses can be used to easily determine which requirements are important for crossing Death Valley without any electronic aids. One manager put it in a nutshell as follows: **"You need agile walkers with good orientation and equipment."** Accordingly, three elements are necessary:

- **Good agility** to be able to cross the desert as quickly as possible.
- **Good orientation,** that is, information about the best route and behavioral rules for crossing the desert in order to reach the destination safely and by the fastest route.
- **Good equipment** such as a compass, a water backpack, sun protection, and footwear in order to guarantee orientation and agility.

Fig. 5 Requirements for crossing Death Valley

1.2 Are there Death Valley scenarios for companies?

I argue that both of these Death Valley scenarios are relevant to companies in current market environments. This is particularly true for companies from highly developed industrialized nations that are not used to surviving in barren market conditions. Let me describe the analogy between Death Valley and the markets in a little more detail.

Many of our markets are gradually developing into hostile Death Valleys. They are overheating due to extreme competition. As a consequence, customers' willingness to pay is becoming as scarce as the rainfall in Death Valley. In the sense of market niches, there is hardly any protective shade to be found.

Accordingly, I define Death Valley markets as *the market sectors in which suppliers are only able to sell their products and services with very low profit margins.*

What is striking about this analogy is that many companies from highly developed industrialized nations are increasingly being drawn into the maelstrom of Death Valley markets. We analyzed the change in market pressure in a scientific survey of managers from 756 companies in 12 industries: 72.4% of all managers believed that their company was under increasing pressure. The following table shows considerable differences between the 12 industries. Pressure in the financial services, telecommunications, and automotive industries is increasing sharply. Although the market pressure perceived in the construction, chemicals, and technology and software industries is less extensive, it is still impacting more than half of the companies surveyed.

	Increasing pressure in the market
Financial services	83.7%
Telecommunications	81.4%
Automotive	79.1%
Consumer & retail	77.3%
Industrial goods & mechanical engineering	77.1%
B2B commerce & distribution	75.8%
Pharmaceuticals & healthcare	74.4%
Energy & supply	73.7%
Transport & logistics	73.7%
Technology & software	66.7%
Chemicals & base materials	65.0%
Others	60.2%
Construction	51.9%
Average	**72.4%**

1.3 Why are Death Valley markets spreading?

The reasons for the spread of Death Valley markets are manifold. I would like to deal briefly with three important factors (see Figure 6): Globalization, digitalization, and crises.

Fig. 6 The three drivers of Death Valley markets

Globalization – enables new cost leaders to force their way onto the markets.

The first of the Death Valley factors is well-known. The profitability of established suppliers is jeopardized because cost leaders from emerging economies force their way onto the markets with low prices. In the 1970s and 1980s, these cost leaders came from Japan, from China in the 1990s and 2000s, and are now increasingly coming from the other BRICS countries such as Brazil, Russia, India, and South Africa as well as various emerging economies such as Indonesia, Iran, Mexico, Nigeria, Pakistan, the Philippines, Turkey, South Korea, and Vietnam.

Due to low wage costs, the costs for manufacturing products are significantly lower there, with the result that suppliers from these countries become the cost leaders in their respective markets. They are therefore able to create momentum in their markets with low prices. Suppliers from highly developed industrialized nations are no longer able to keep up in terms of prices.

Meanwhile, the profitability of established suppliers is additionally in jeopardy because the quality offered by the new cost leaders is also becoming increasingly competitive. Although the quality improvement process of competitors from emerging economies takes time, it is inexorable. Up-and-coming cost leaders are already finding it increasingly easy to keep pace in terms of quality in industrial goods markets as well. Above all, Chinese suppliers have long since ascended to the top of the quality ladder in a number of mechanical engineering segments, for example.

Digitalization – creates transparency and reduces shadowy protected areas in markets.

The second driving force, digitalization, plays a critical role in the occurrence of Death Valley markets. In particular, digitalization of the product search and purchase process gives rise to market transparency that simply did not exist before. This causes the customers' willingness to pay to decline to such an extent that it is of little attraction any longer to companies from high-wage countries.

At its core, this growing market transparency is being driven by the increasingly extensive options available in online channels. They are making it increasingly easy to compare products and services in terms of both a) the price and b) the benefit components of the offers.

a) **Price transparency:** Platforms, online shops, and comparison portals present a quick overview of the prices of comparable product and service offers in an increasingly uncomplicated manner.

b) **Performance transparency:** Online visualization tools enable a considerably more extensive comparison of the characteristics of even very complex products. They make it mercilessly clear which product specifications are to be had from which supplier at the lowest price.

Players such as *Alibaba, Amazon Business, Direct Industry, Europages, rajapack,* and *Mercateo* are becoming increasingly dominant in commerce. Online retailers originally focused on B2C, such as *Amazon*, are pushing forward into ever more product dimensions. These also increasingly include classic B2B product segments – from electronic accessories and construction site materials to more complex machines. *Amazon Business* has long since been battling with *Alibaba* for global market leadership in the area of B2B purchasing. The European players are still trailing behind.

The online sales channel is additionally ratcheting up globalized competitive pressure by providing international competitors with inexpensive market access. As a result, pressure on profit margins increases even further. Price competition has become more extreme in all of the market sectors in which this online comparability has been established.

Crises – act as catalysts for Death Valley markets.

In combination with the first two Death Valley factors, crises have the same effect as fire accelerants. In many industries, price pressure was already ratcheted up by the financial crisis back in 2008. In turn, the coronavirus pandemic has significantly speeded up the trend toward digitalization. Company weaknesses that were able to remain concealed during the years of economic growth prior to the pandemic have since been mercilessly revealed.[2]

A particular dynamism therefore arises from the interaction of all three factors. Since globalization, digitalization, and crises have occurred simultaneously over the past two decades, the development of Death Valley markets has accelerated rapidly.

If it is true that Death Valley conditions are now increasingly prevailing in many markets, companies will be forced to deal with both of the scenarios from our mental experiment (see Figure 7). This means that companies will also have to fulfill one of the aforementioned requirements and

- either become perfectly adapted minimalists in order to survive in Death Valley in the long term,
- or become agile walkers in order to escape Death Valley.

I deal with both options in the following.

[2] To obtain a deeper understanding of (sales)strategies in times of crisis, we conducted a study in our department: Wieseke, Schmitz, Knoess, Kassemeier (2020), *(Sales) Strategies in an Age of Coronavirus and Recession – Insights and Findings from 295 Sales Professionals,* Sales Management Department, Ruhr University of Bochum.

Fig. 7 Survival strategies in Death Valley markets

1.4 How is long-term survival possible in Death Valley markets?

The world of flora and fauna impressively shows how survival in Death Valley is possible despite the hostile conditions. The most important survival strategy involves optimal adaptation to the dry, hot environment. Specifically, this means maximum **efficiency** in order to consume as little water as possible!

The **Atriplex hymenelytra** is a prime example of efficiency (see Figure 8). As can be seen from the light color of its leaves, its surface is coated with salt. This leads to an efficiency effect because the lightness of the salt causes more sunlight to be reflected, and thus reduces water evaporation. The **Atriplex hymenelytra** belongs to the "orache" plant genus, and so to the family of so-called foxtail plants. These plants combine the reflective effect on the surface of their leaves with wide-ranging root systems that can grow to lengths of up to 15 meters in Death Valley.

Fig. 8 The efficiency of the world of flora in Death Valley

Prairie grass is a wonderful example of such root systems. The figure shows large distances between the tufts of grass. This is because extensively branching root networks also extend around the tufts of grass just beneath the surface of the desert floor. As a result, the prairie grass offers the optimal surface for absorbing the scant rain that falls in Death Valley.

Besides plants, animals also survive in Death Valley. These include reptiles (such as snakes, geckos, and lizards) and arthropods (such as insects, beetles, and spiders). Their common feature is maximum efficiency due, among other aspects, to their poikilothermic nature, which adjusts their body temperature to the ambient temperature.

Efficiency strategies in Death Valley markets. As in the ecological Death Valley environment, maximum efficiency is needed to survive in economic Death Valleys. Comparable to the energy-saving mode of plants and animals, only companies that are perfectly tailored to low costs can survive in Death Valley markets. These cost leaders operate extremely sparingly with the available resources and implement lean internal structures. Like the perfectly adapted Atriplex hymenelytra, these companies actually thrive under such circumstances.

Companies that are not among the price and cost leaders perish in Death Valley markets. It is impossible for the majority of companies from highly developed industrialized nations to implement an "Atriplex strategy." The wage costs alone represent an insurmountable obstacle (see Figure 9).

In order to survive the prevailing price competition, one cost-cutting program usually follows the next in such companies, and funds for investments become increasingly scant. As in the real-life Death Valley, their demise is therefore merely a question of time.

Fig. 9 Survival strategies in Death Valley

This therefore raises the question of how to escape from Death Valley markets, which takes us to the second scenario of the above mental experiment.

1.5 Searching for green hill markets – opportunities through solution business models

The second scenario of the mental experiment involved crossing Death Valley in order to reach more fruitful regions. I call these regions "green hills." Green hills are landscapes in which there is sufficient food and water to survive.

There are green hills on our markets as well. This is where companies exist that have succeeded in evading digital transparency and the price pressure of global competitors. Although they are not among the price/cost leaders in their industry, they are positioned very profitably.

The core characteristic of the green hills is a high willingness of the customer to pay. Accordingly, I define green hill markets as *market sectors in which customers are willing to pay for a company's products and services at a price that enables the supplier to achieve a healthy profit margin.*

Intelligent business models have to be developed in order to achieve this willingness to pay. Companies that have successfully made their way into green hills have systematically determined how they can enhance the benefit for their customers in order to increase their willingness to pay. They can describe unequivocally and clearly which benefit is relevant to the customers, leading to an increase in their willingness to pay.

In the majority of cases, it is necessary to shift business models in the direction of **solution business models** in order to achieve this significantly higher customer benefit. I consciously define "solutions" broadly here as all approaches in which customers are offered an additional benefit through more or less complex solutions in order to increase the individual profitability of the customer companies.

Very often, services that are intended to lead to a new added value are established around tangible products. Solution business models are aimed at helping customers to optimize their value chains. For instance, suppliers take over entire customer production process steps or guarantee the availability of products instead of selling them.

The logic is therefore that the solutions are used to provide customers with an additional benefit, as a result of which both the customers' willingness to pay and the supplier's profit margins are increased. Added to this is the fact that solutions can become barriers to market entry. The increased complexity of the solutions compared to classic products makes it difficult for the competition to copy the offer with the same quality. Complex solutions also help the supplier to evade online platforms, and thus price comparison transparency.

Fig. 10 Green hill markets

The trend toward solutions is unstoppable because it goes hand-in-hand with the development of digital technologies. Even – and especially – companies that actually started out in the pure product business are now offering customers solutions surrounding their products. This mega trend is making its presence felt to such an extent that only 43% of the 845 B2B companies that we analyzed still regard themselves as being in the pure product business. Sums in the high billions are being invested in new solution business models in diverse industries, from IT security suppliers and manufacturers of paper machines to commodity chemicals suppliers.

A range of books about the strategic background to these business models is now available. However, all of the solution approaches are based on the same logic, irrespective of whether they are dubbed "complex digital solutions," "servitization," "Industry 4.0," "value added services," or "complex service solutions:" Suppliers are establishing new benefit dimensions by optimizing their customers' value creation. In Chapters 4–7, I deal with these business models in more detail and provide numerous examples.

Let us sum up. I am making a total of three statements:

1. The majority of our markets are gradually developing into hostile Death Valleys with a low willingness of the customer to pay.
2. In order to survive, (almost all) companies from highly developed industrialized nations have to find their way to the green hill markets, where customers are more willing to pay.
3. To achieve this objective, companies must develop new solution business models that enhance the customer benefit in such a way that the customers' willingness to pay increases.

Figuratively speaking, this involves escaping the maelstrom of Death Valley markets time and again and shifting toward green hill markets (see Figure 11). To do this, we need a compass that helps us to come up with successful solution business models again and again.

Fig. 11 The road from Death Valley markets to green hill markets

WHY SALESTEGY?

STRATEGY – FINDING THE RIGHT BUSINESS MODELS

PRACTICAL EXAMPLES

SALES – IMPLEMENTING NEW BUSINESS MODELS IN SALES

Chapter 2
The concept of SALESTEGY

Message of this chapter

This chapter deals with the concept of SALESTEGY – the combination of SALES & STRATEGY. SALESTEGY is necessary to find the right solution business models and implement them successfully.

What should you have learned by the end of the chapter?

- Solution business models necessitate the combination of the sales function and strategic management (SALESTEGY). This applies both to finding promising business models and their subsequent implementation.

- Five messages are relevant in this case:
  - › No. 1: SALESTEGY is necessary to reach green hill markets.
  - › No. 2: SALESTEGY means using sales in a dual role:
    a) as a catalyst when finding new solution business models and
    b) as an implementer when introducing them.
  - › No. 3: The more complex the business models become, the more policy competence sales should be given. Solution champions have established sales at board level.
  - › No. 4: The importance of personal selling increases along with the importance of solution business models.
  - › No. 5: The shortage of skilled staff and managers in sales is becoming a growth bottleneck in B2B companies.

Let us return to the last figure from Chapter 1. It shows that the way out of the Death Valley markets to the green hill markets is made possible by solution business models. Solution business models offer companies – particularly for established companies from highly developed industrialized nations – a major opportunity to maintain their competitive advantage.

Nevertheless, the road to establishing solution business models is rocky. Aside from the readily told tales of success, numerous companies have failed on the road to solutions. According to our estimates, around half of all solution strategies actually fail to meet management expectations to such an extent that they are classified as failures.

The biggest challenges lie in two areas (see Figure 12), namely

1. finding the right solution business models, and
2. successfully implementing these business models.

Both carry their own particular risks:

Risk no. 1: Losing one's bearings. It is possible to lose your bearings because the number of possible business models is enormous. Chapter 3 deals with this wealth of possibilities in the following. There is a risk of becoming bogged down in the variety of options. As a result, I subdivide this variety into five steps

(Chapters 4 and 5) and determine the competitive advantage (Chapter 6).

Risk no. 2: Failing during implementation. The second risk lies in implementation. According to our research, implementation often fails in sales. The biggest risk factor here is in structuring and managing sales in the conventional manner. In Chapters 8–14, I therefore show which adaptations are necessary in sales in order to successfully implement solution business models.

My core statement in this chapter is as follows: The combination of SALES & STRATEGY is crucial in order to find the right solution business models and implement them successfully. I call this combination SALESTEGY.

Fig. 12 Challenges on the road to green hill markets

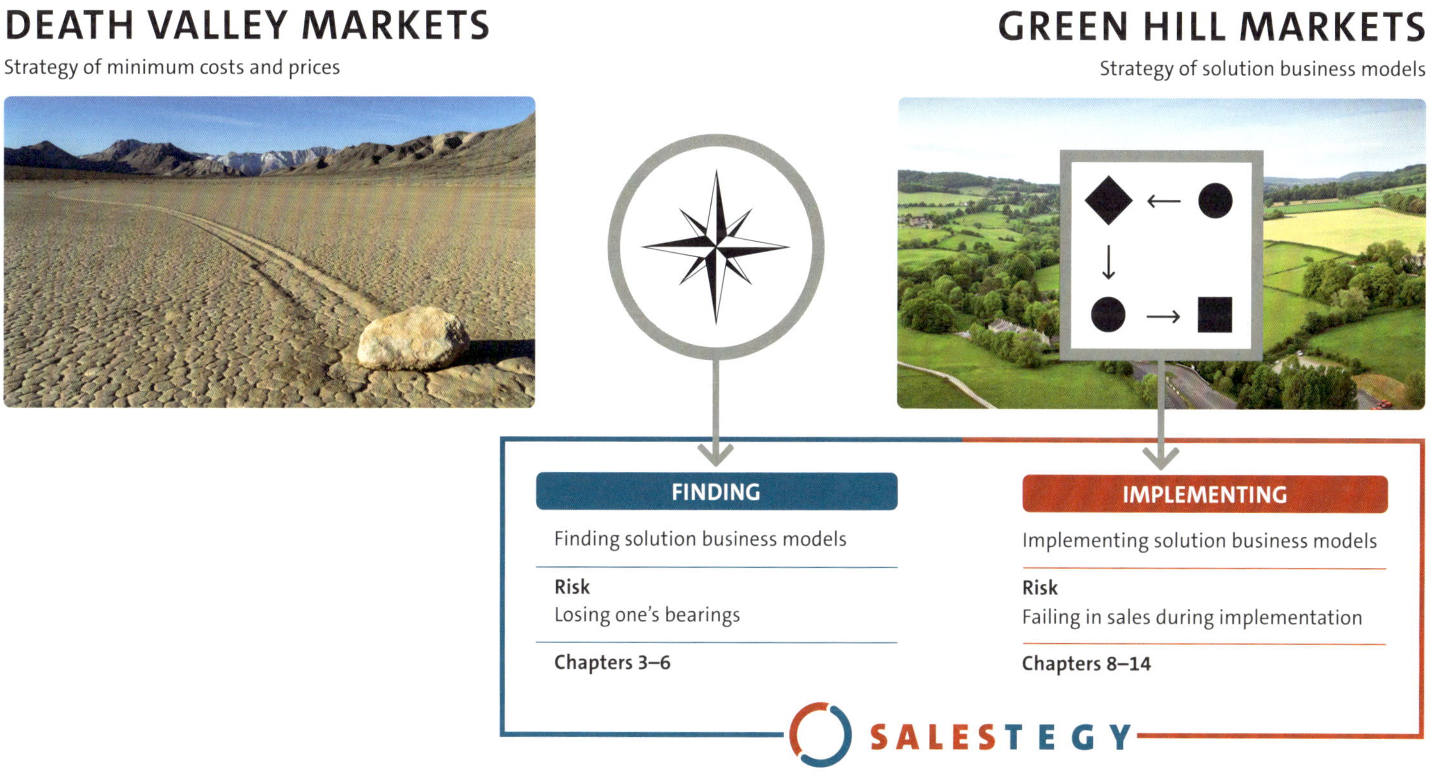

The concept of SALESTEGY runs counter to the classic distribution of roles that is still practiced in many companies. It is shown in the left-hand area of the following table. According to this, strategic management and sales act remotely from one another. There are no regular coordination meetings. The distribution of power is such that the sales department does what it is told by the strategic management.

Conversely, the SALESTEGY role distribution provides for the close intermeshing of strategic management and sales. This is shown in the right-hand column of the table. Here, the sales department acts on an equal footing, since sales department representatives are involved in strategic decisions or are even permanently integrated at board level.

	Role distribution between strategic management (STM) & sales			
	Classic		SALESTEGY	
	STM	Sales	STM	Sales
Combination of STRATEGY & SALES	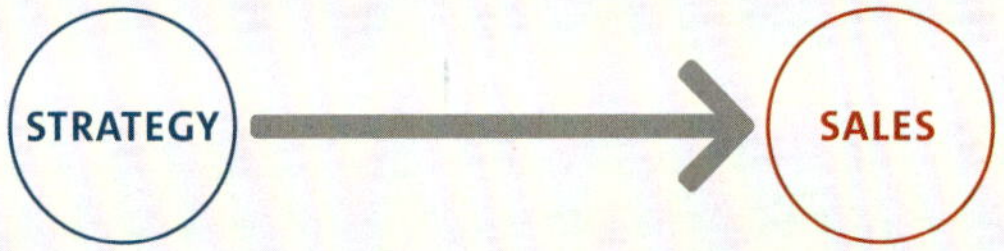		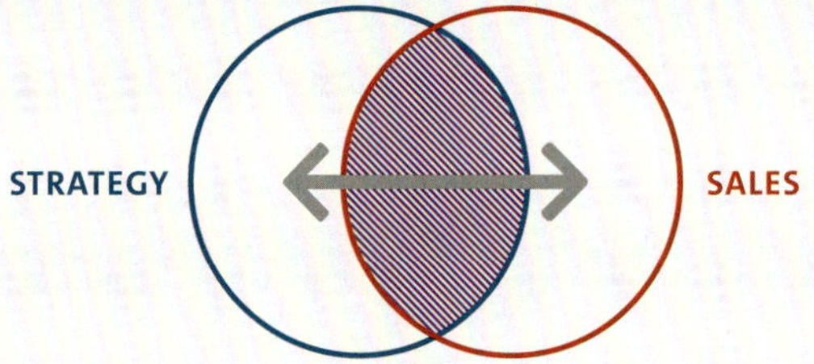	
Distribution of power – who determines	STM determines	Sales executes	Sales is represented in STM	

The next table shows the distribution of roles when finding and implementing new business models. Again, the classic distribution of roles is shown on the left. There, strategic management is responsible for finding new business models, whereas sales concentrates on implementation. The management board conducts feedback loops primarily with technical product development and occasionally also with marketing, but rarely with sales.

In the SALESTEGY role distribution, strategic management and sales cooperate closely with one another, both in finding and implementing new business models. Sales is used virtually as a market sensor in the search for promising business models. Since no other company function operates so close to customers, sales is best able to assess the benefit to customers and their willingness to pay as well as feasibility. Many of the 18 successful solution business models that I introduce in Chapter 7 originate from sales team ideas.

I have consciously used the picture of the compass above because it is needed to escape from the maelstrom of Death Valley markets and move toward green hill markets. Since we are dealing with the search for business models, this involves a *strategic compass*. I believe that the sales department's resources are suitable for adjusting this strategic compass and avoiding serious misjudgments.

In implementing successful solution business models, good companies rely on continuous feedback loops between strategic management and sales. In the case of failing business models, however, we often found that the sales department was left to its own devices and that board-level intervention only occurred when implementation failed to succeed, leading to more pressure being applied.

Role distribution in new business models

	Classic		SALESTEGY	
	STM	Sales	STM	Sales
Finding new business models	✓		✓	✓ Market sensor
Implementing new business models		✓	✓ Feedback loops	✓

What SALESTEGY does *not* mean is that sales takes over the main tasks of strategic management and vice versa. Strategic management is also mainly responsible for finding new business models in the case of SALESTEGY, whereas sales is mainly responsible for implementing them (see Figure 13). However, SALESTEGY means that strategic management cooperates closely with the sales function in finding new business models and supports it in implementing them. Both work closely together during the finding and introductory phases of new business models.

Fig. 13 The distribution of roles between SALES & STRATEGY in the SALESTEGY approach

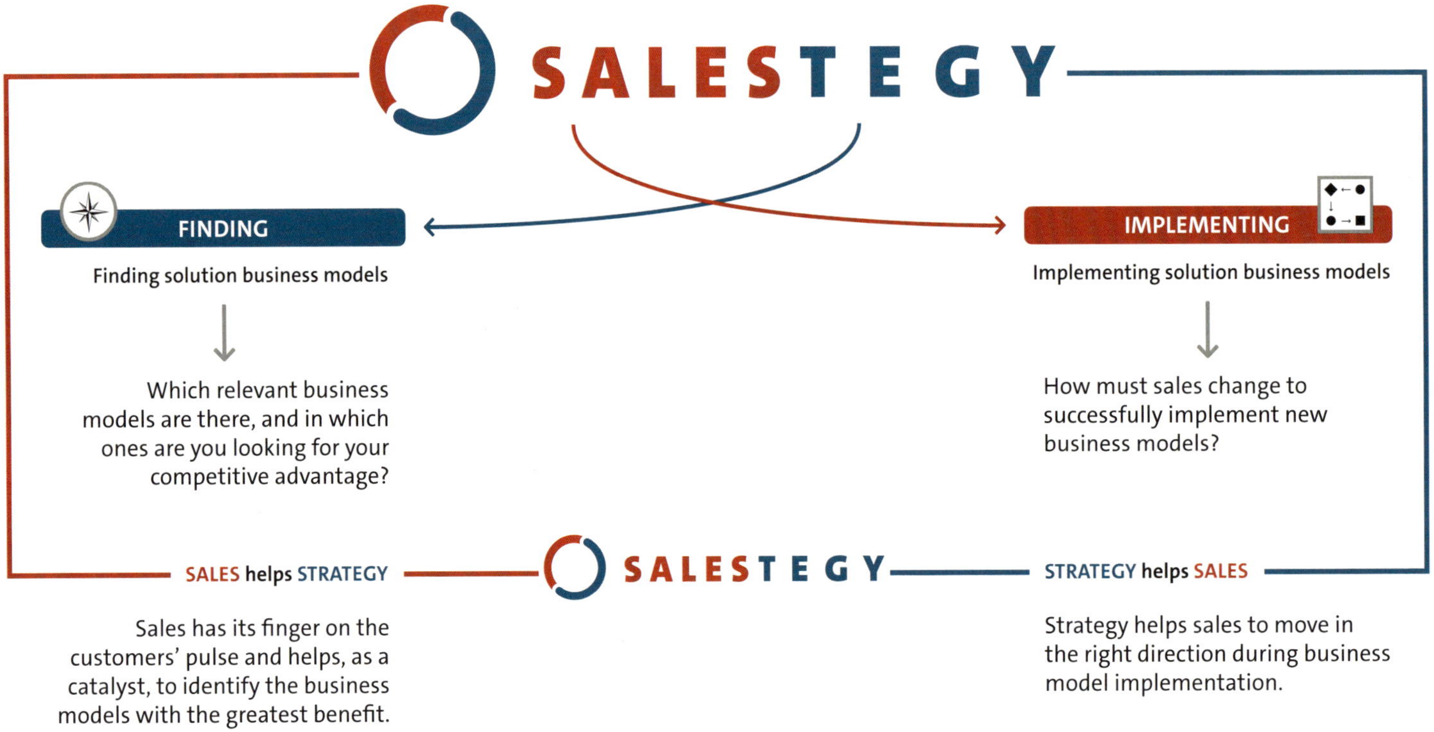

2.1 Five messages concerning SALESTEGY – the combination of SALES & STRATEGY

Based on the previous considerations, there are five central messages as regards SALESTEGY:

No. 1: SALESTEGY is necessary to reach green hill markets.

No. 2: SALESTEGY means using sales in a dual role: a) As a catalyst when finding new solution business models and b) as an implementer when introducing them.

No. 3: The more complex the business models become, the more policy competence sales should be given. Solution champions have established sales at board level.

No. 4: The importance of personal selling increases along with the importance of solution business models.

No. 5: The shortage of skilled staff and managers in sales is becoming a growth bottleneck in B2B companies.

Message no. 1
SALESTEGY is necessary to reach green hill markets.

The first message picks up on what I have described so far regarding the link between the Death Valley scenarios and the SALESTEGY concept. Let us compare the two scenarios from Chapter 1 once again (see adjacent table).

- Minimal costs are the core requirement for long-term survival in Death Valley markets. SALESTEGY is *not* necessary here because everything is concerned with optimizing the purchase price. It involves the sale of comparatively simple products and services that require little explanation and can therefore be distributed primarily via (online) retailers.
- Conversely, SALESTEGY, or the close combination of SALES & STRATEGY, is necessary in order to find and successfully implement green hill business models. This is related to the complex search for solution business models with which the profitability of the customer companies can best be increased. Selling these business models is also complex and requires direct feedback from SALES & STRATEGY during their introduction.

	Long-term survival in Death Valley markets	Search for green hill markets
Requirements for companies	Minimum costs	Maximum benefit
Customer reason for purchasing	Minimum purchase price	Customer profitability increases
Is sales becoming a bottleneck?	NO	YES
	↓	↓
SALESTEGY relevant?	NO	YES

Message no. 2
SALESTEGY means using sales in a dual role:
a) As a catalyst when finding new solution business models and b) as an implementer when introducing them.

SALESTEGY means using sales in a dual role on the road to new business models. The sales department is used as a market sensor and a catalyst for the new business models and, at the same time, as an implementer in the direction of customers (see Figure 14).

The salesforce as a catalyst for new business models. Due to its proximity to customers, the sales department should always be involved in business model analyses. This especially applies to complex B2B markets. Sales employees are often the only ones who are able to solve customers' problems, as they are in constant contact with the customers and are familiar with their current challenges.

The sales department's knowledge of *what* can be marketed, on *whom* the focus should be placed among customers, and *how* the willingness of customers to pay can be achieved in view of the competitors on the market is second to none. In the complex solutions sector, the search for answers to these questions cannot be delegated to external market research agencies.

However, this sales department knowledge is not only relevant for radical business model changes. It is also essential when gradually adapting a business model – when extending customer target groups or adapting the characteristics of products and services to current market trends, for example.

Without a feel for customers' business models, it is hardly possible to develop viable innovations in B2B markets. There is no other corporate function other than sales in which this market sensor is so extensively pronounced. This sales sensor function becomes a bottleneck, particularly in fiercely contested B2B markets, meaning that the sales department gains in strategic importance.

The salesforce as the implementer of new business models. The second function of the sales department is in implementing new business models. If the new business models involve complex solutions, the salesforce is the most important implementer. If only one of the six factors of sales (sales employee skills, sales compensation, sales leadership, lead management, sales structures, sales methods, see Chapters 9–14) is not right, the solution business model fails.

Fig. 14 The dual role of sales

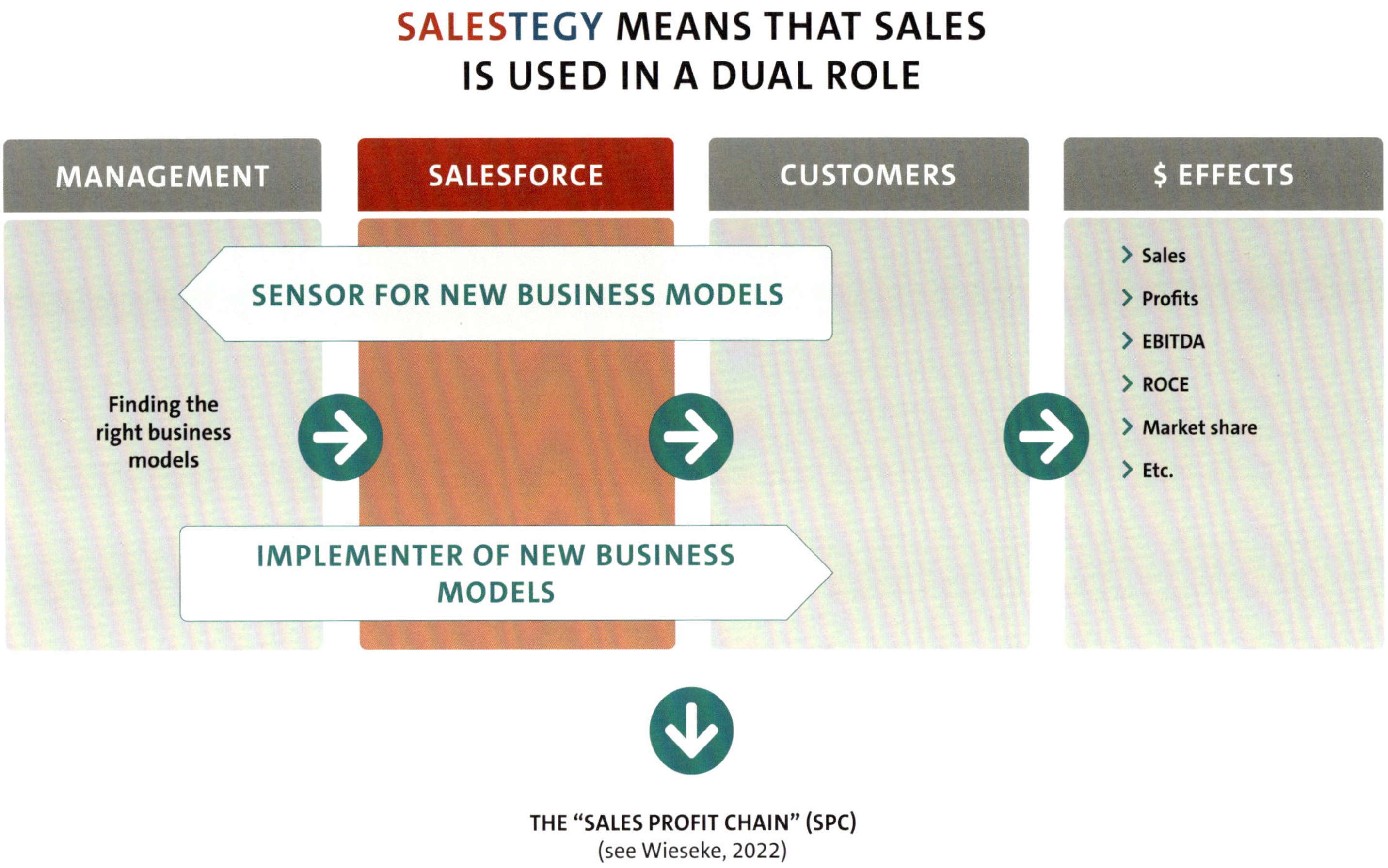

Message no. 3
The more complex the business models become, the more policy competence sales should be given. Solution champions have established sales at board level.

In current management textbooks, sales is portrayed merely as an executive entity. However, this approach no longer meets the requirements of today's market environments.

If companies want to move into solution business models, and therefore into green hill markets successfully, sales management must be on an equal footing with strategic management. Managers with sales expertise must be involved in setting the course of corporate strategy. This refers to managers who have themselves worked in sales for a long time and currently have a good relationship with customers and the salesforce.

Our observation from numerous studies and projects with companies is that the past two decades have already shown a gradual upgrading of the sales management. According to this, a distinction can often be made between the following three phases over time (see Figure 15).

Phase 1: The sales management is subordinate to the strategic management.

- Product-centric company
- **Role distribution:** The classic distribution of roles in product-centric companies is such that the executive level issues specifications to sales. Innovative products are created in research and development departments, manufactured in production, and brought to customers by sales. In the hierarchical and temporal sequence, sales is therefore far at the back.
- **Hierarchy:** The top-ranking sales managers are subordinate to business unit executives here.
- **Classic quotations:** *"We have a clear hierarchy here: The gentlemen from research and development, the men from production, and the people from sales." – "Sales is the most operational of all corporate functions." – "Here, sales is a cost block that we have to cap." – "We've often talked about outsourcing sales."*

Phase 2: The sales management moves closer to the strategic management.

- Companies on the way to becoming solution providers
- **Role distribution:** The strategic management delivers the business models. The sales management delivers the knowledge about how sales can be controlled and customers acquired to achieve this. Sales is recognized as a catalyst and a veto entity in the case of innovations because its knowledge of the market is most extensive.
- **Hierarchy:** The top-ranking sales managers are business unit executives.
- **Classic quotations:** *"Our sales department can't be outsourced." – "We have to continue investing heavily in sales to be able to grow." – "Sales is the driving force of our innovation."*

Phase 3: The sales management is integrated in the strategic management.

- Solution champions
- **Role distribution:** Strategic management and sales management develop the thrust of the future business models together.
- **Hierarchy:** At least one sales manager is a fixed part of the company's management board, as the Chief Sales Officer (CSO), for instance.
- **Classic quotations:** *"We are a sales-driven company." – "We don't allow just anybody to visit our customers' decision-makers. You rise through the ranks to sales!"*

Fig. 15 The distribution of power between strategic management and sales

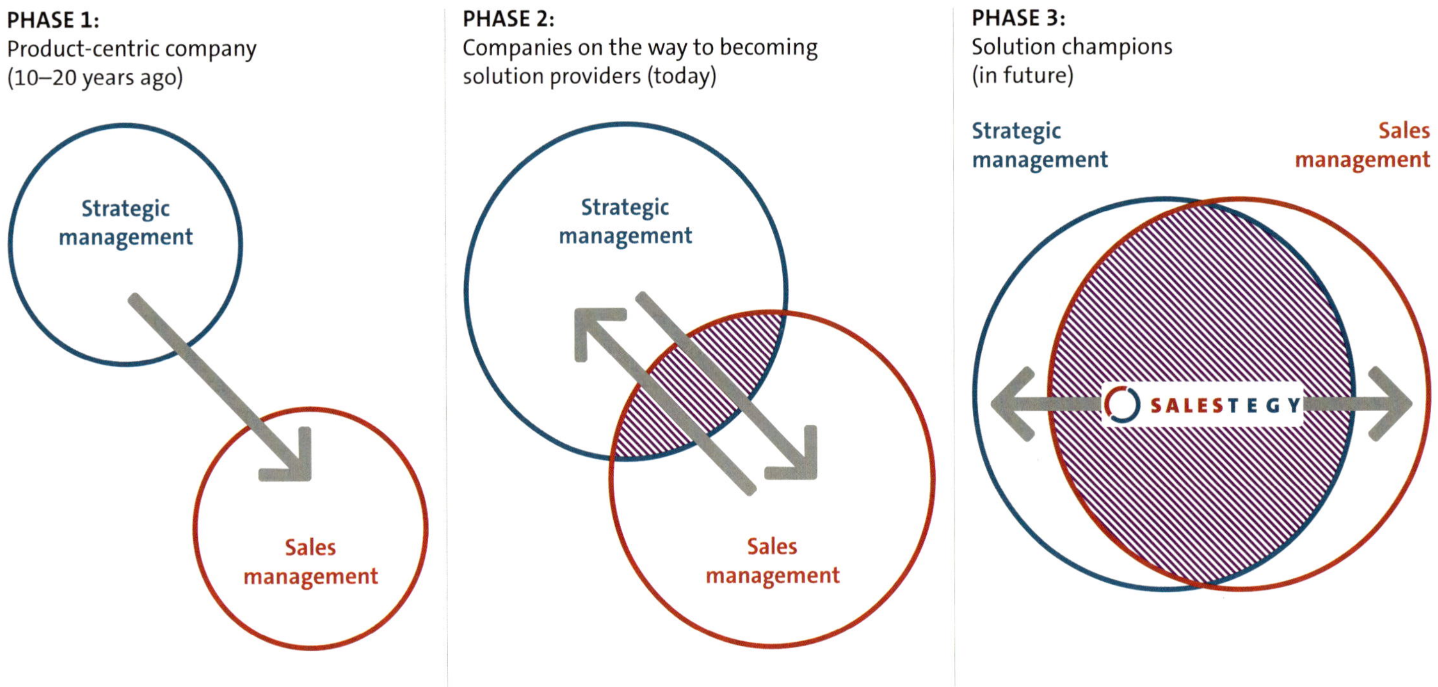

Message no. 4
The importance of personal selling increases along with the importance of solution business models.

When I talk about "sales" in this book, I am specifically referring to those forms of sales that function through personal selling. Personal interaction is crucial in this, irrespective of whether it takes place through a classic field salesforce visit or a video conference.

But what makes personal selling so important? We move closer to the answer when we take a look at the aforementioned failure rate of new solution business models. One very significant reason for such failures lies in sales. It is the central bottleneck.

This is due to two major obstacles to purchasing that have to be overcome in the case of solution business models (see Figure 16). Sales becomes a bottleneck because only it is able to break down these purchasing obstacles. Only outstanding sales employees who are led by outstanding sales managers are able to do this.

Fig. 16 Two purchasing obstacles in solution business models

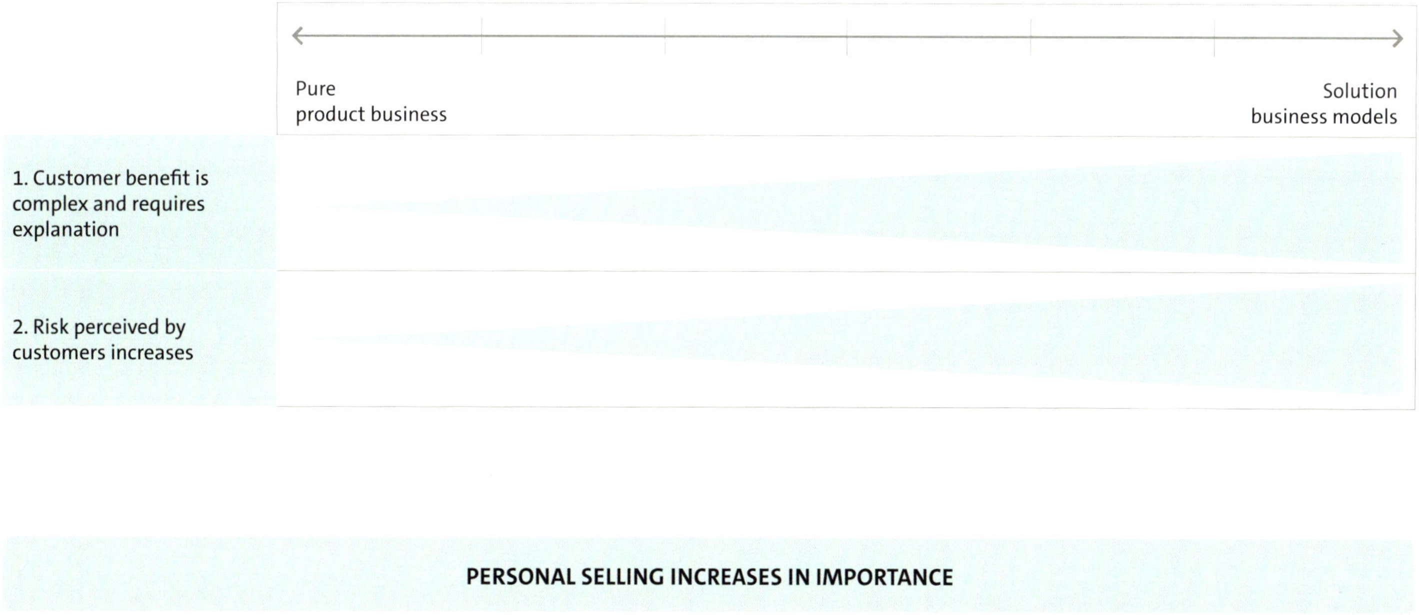

First obstacle to purchasing solutions: The customer benefit is complex and requires explanation. The majority of solutions are so complex that they have to be explained to the respective customer. Accordingly, they have to be sold using personal contact. For a foreseeable time, no other channel will be able to convey the complexity resulting from the customized offer for solutions. Added to this is the fact that the development of complex solutions usually involves multiple feedback loops between suppliers and customers. Various functions such as purchasing, sales, research and development, production, and service are usually involved on both sides. Sales acts as the central coordinator in this.

The requirements for sales managers and employees increase along with the complexity of the solution business models. However, many sales employees in B2B industries have a pure product sales background, in which "character selling" was the clincher. *"Our machine manages 10,000 capsules per hour, offers an LED safety scanner, and only consumes 500 kWh."* But this approach is doomed to fail in solution selling. As customers find it more difficult to imagine the (non-visible) benefit of complex solutions, this results in the need for outstanding sales employees who are cable of "benefit selling."

Second obstacle to purchasing solutions: The risk perceived by customers increases. Each new solution initially involves risks for the customer. The following quotation conveys this vividly:

"How dependent will this supposedly comfortable solution make me? How dependent will I be, for example, if the supplier actually takes over the machines for one of our process steps using its own employees here at our plant? What is its intention if it no longer sells the machine to me but offers to lease it? What long-term cost risks will I face? What does it mean for my accounting? Can I rely on the fact that the supplier will really provide a replacement if there are problems on the machine?" (COO of a producer of systems for foodstuffs).

Risk therefore plays a central role in purchasing solutions. Understanding and addressing this customer risk is demanding because there are many types of risk, and the individual *perception* of the individual customer is ultimately what is important. In order to reduce this feeling of risk, the sales department has to have a close relationship with the customer. Possible misgivings can only be cleared up and the customer's withdrawal prevented through personal selling.

Message no. 5
The shortage of skilled staff and managers in sales is becoming a growth bottleneck in B2B companies.

The increased requirements for sales managers and employees in the solution business raise the question of whether the labor market offers a sufficient pool of qualified specialists and managers.

The *Talent Shortage Survey* paints a bleak picture in this regard. As part of this broad-based study, the *ManpowerGroup* surveyed 41,700 corporate recruiters in 42 countries.[3] According to this, sales was in second place when respondents were asked which were the "hardest jobs to fill." Tradesmen were in first place, followed by engineers and technicians in third and fourth place, respectively.

So the quantity of young sales talents is in a bad way. But what about the quality? Can sales employees and managers meet the increased requirements for sales work, particularly in the complex B2B business?

To find out, we conducted our own study in 12 industries, surveying 510 managers in the process.[4] The results of our study reveal clear deficits in newly recruited sales employees. Accordingly, 75% of new sales employees fail to adequately meet expectations.

Although *technical-functional skills* do not constitute a bottleneck (see Figure 17), with only 1.4% of respondents seeing relevant deficits among sales employees, we did discover major deficits in the skills that are important for the solution business. The top three skill deficiencies among sales employees are *strategic actions and decision-making (67%), creative problem-solving (63%), and customer analysis and customer understanding (58%).*

However, it is precisely these skills that are critical in the solution business. One of the survey participants from the telecommunications industry described this dilemma as follows: *"We need more solution business, but we're finding ever fewer sales employees for precisely that."*

Put simply, this means that there will be a massive shortage of skilled workers where the requirements for highly qualified sales employees are greatest. This is precisely where B2B companies are currently looking to find their competitive advantage: in business involving complex solutions. This means that sales is becoming a bottleneck for the growth of B2B companies.

In Chapter 9, I return to the issue of the necessary role requirements in sales and deal with the profiles that are needed for the new digital solution business models.

---

3 ManpowerGroup (2020), *Talent Shortage Survey: Closing the Skills Gap: What Workers Want*, p. 3.

4 Wieseke, Schmitz (2015), *Focus Study Concerning Skill Requirements in Sales*, Sales Management Department, Ruhr University of Bochum

Fig. 17 Study results regarding skill deficits among sales employees

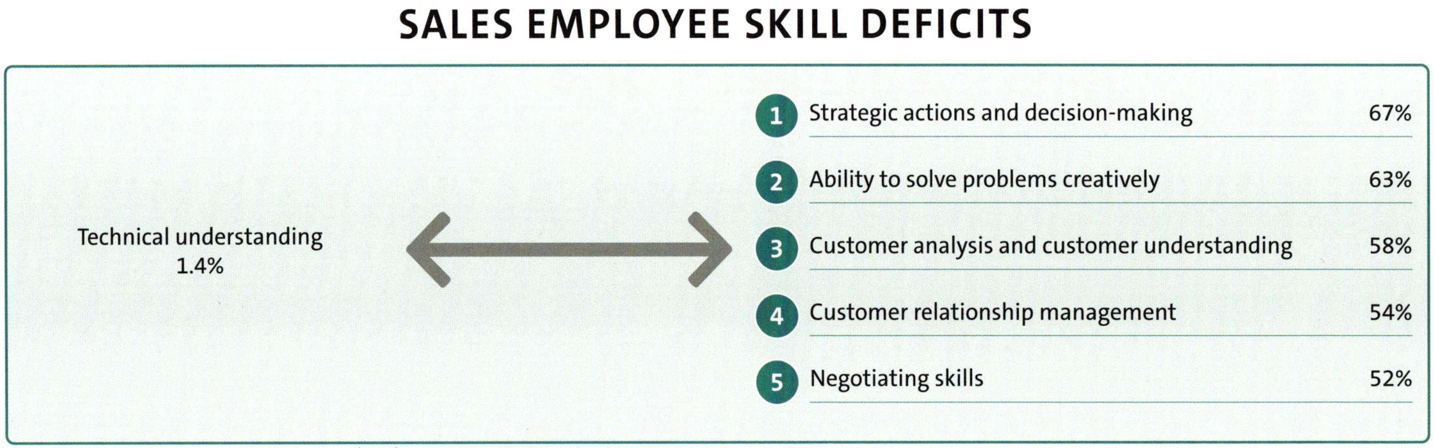

Source: Sales Management Department, Bochum (2015); n=510 managers in 12 industries

2.2 SALESTEGY as the basis of the chapters in this book

Figure 18 shows the sub-tasks of SALESTEGY. These form the basis for the coming chapters because they involve the steps that are necessary for *finding* and *implementing* new business models. To find new business models, it is helpful to the supplier

1. to reduce the complexity of the many possible business models (see Chapter 3),
2. to distinguish between five business model steps (see Chapter 4),
3. to use the so-called SNOOFY factors to differentiate business model steps (see Chapter 5), and
4. to check where the greatest competitive advantage can be achieved by conducting a business model audit (see Chapter 6).

When implementing new business models, there are six major areas where Death Valley risks are lurking in sales. These are:

1. Sales employee skills (see Chapter 9)
2. Sales compensation (see Chapter 10)
3. Sales leadership (see Chapter 11)
4. Lead management (see Chapter 12)
5. Sales structures (see Chapter 13)
6. Sales methods (see Chapter 14)

Fig. 18 Overview: How the SALESTEGY approach is reflected in the following chapters

SALESTEGY

SALES helps STRATEGY

Sales has its finger on the customers' pulse and helps, as a catalyst, to identify the business models with the greatest benefit.

STRATEGY helps SALES

Strategy helps sales to move in the right direction during business model implementation.

STRATEGY

Main task: **Finding new business models**

Sub-tasks:

1. Reducing complexity — Chapter 3
2. Distinguishing between five business model steps — Chapter 4
3. Using SNOOFY factors for differentiation — Chapter 5
4. Business model audit – identifying market opportunities — Chapter 6

SALES

Main task: **Successfully implementing new business models**

Sub-tasks:

1. Adapting sales employee skills — Chapter 9
2. Adapting sales compensation — Chapter 10
3. Adapting sales leadership — Chapter 11
4. Adapting lead management — Chapter 12
5. Adapting sales structures — Chapter 13
6. Adapting sales methods — Chapter 14

The second part of the book deals with the **STRATEGY** area of **SALESTEGY**. Transferred to the imagery of nature, the following Chapters 3 to 6 concerning the finding of new business models can be portrayed as in Figure 19.

Chapter 3 involves finding a logic that we can use to reduce the variety of possible business models to a manageable level. In Chapter 4, I show that we can use this logic to distinguish between five business model steps (commodity, premium, 1st-step solutions, 2nd-step solutions, super-complex solutions). Chapter 5 describes the SNOOFY factors (***S****ervitization,* ***N****urture customer processes,* ***O****perate processes for customers,* ***O****utcome-based pricing,* ***F****ar-reaching relationship, and* ***Y****ield customer value-in-use*), which we can use to determine the step of a business model. In Chapter 6, I describe the SNOOFY business model audit, with which the competitive advantages in the five business model steps can be identified.

Fig. 19 Chapter overview – Part 2

DEATH VALLEY MARKETS

Strategy of minimum costs and prices

GREEN HILL MARKETS

Strategy of solution business models

FINDING

Finding solution business models

CHAPTER 3

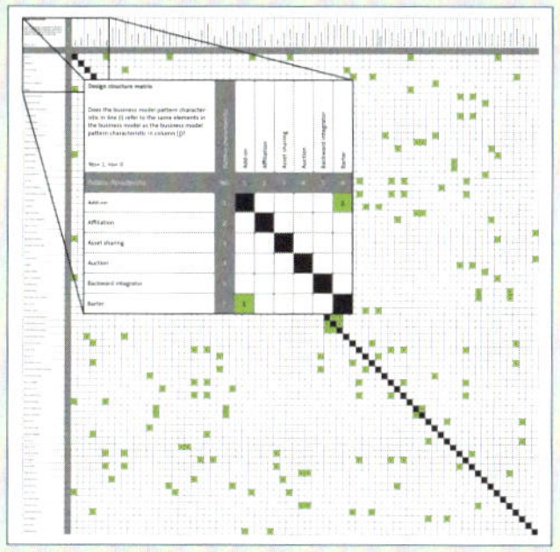

Reducing complexity

CHAPTER 4

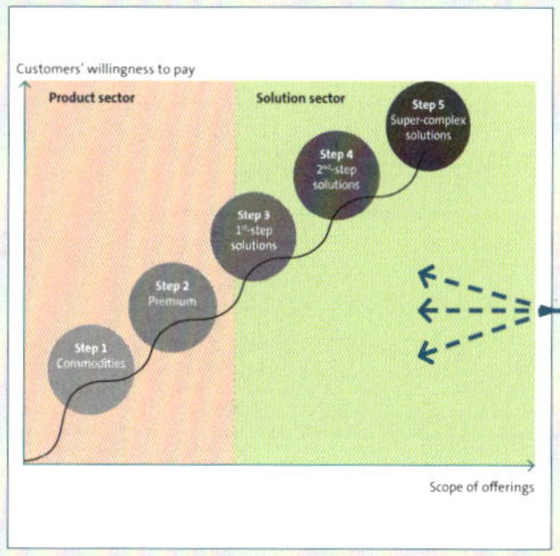

Distinguishing between five business model steps

CHAPTER 5

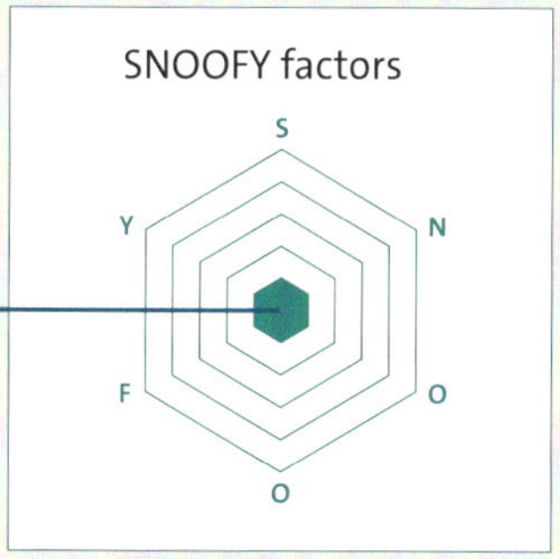

Using SNOOFY factors for differentiation

CHAPTER 6

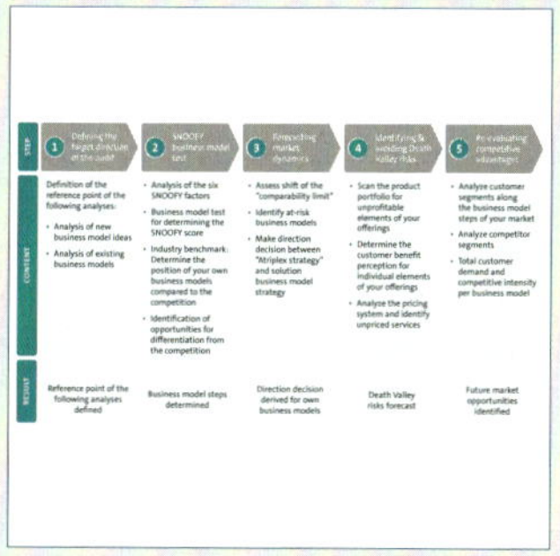

Audit: Identifying opportunities for competitive advantages in the five steps

WHY SALESTEGY?

STRATEGY – FINDING THE RIGHT BUSINESS MODELS

PRACTICAL EXAMPLES

SALES – IMPLEMENTING NEW BUSINESS MODELS IN SALES

Chapter 3

1,000 business model options: Can't we make it simpler?

Message of this chapter

A variety of B2B business models are possible in the digital age. They can be broken down into a few business model steps.

What should you have learned by the end of the chapter?

- In our research, we determined 74 B2B business model patterns that are possible in the digital age of Industry 4.0.
- We can subdivide this variety into business model steps, which are oriented toward two dimensions:

  1. Customers' willingness to pay
  2. The scope of offerings

DEATH VALLEY MARKETS

Strategy of minimum costs and prices

GREEN HILL MARKETS

Strategy of solution business models

FINDING

Finding solution business models

CHAPTER 3

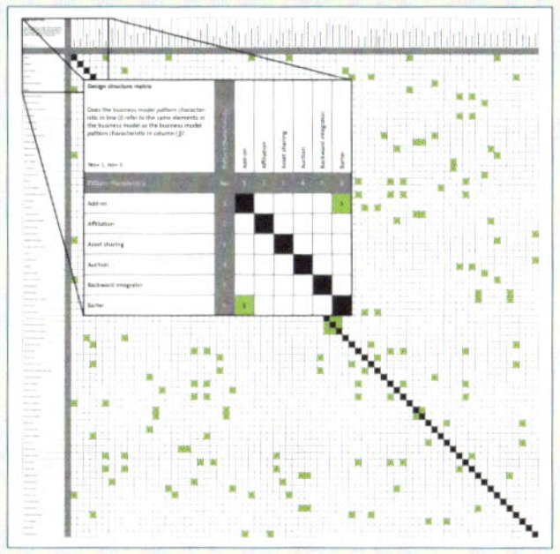

Reducing complexity

CHAPTER 4

Distinguishing between five business model steps

CHAPTER 5

Using SNOOFY factors for differentiation

CHAPTER 6

Audit: Identifying opportunities for competitive advantages in the five steps

The following chapters deal with the business models that are conceivable in your target markets. You will only be able to derive the potentials for your competitive advantage if you are familiar with their positioning.

The options for creating solutions for customers appear boundless in the digital age. However, this wealth of options poses a risk that can take us back to Death Valley. What happens if our customers are not yet ready for our new solutions? What if we are not at all in a position to sell the solutions with our current sales structure and our current sales team?

This gives rise to the urgent question of which of all the possible new business model options we should drive forward next. Acting without orientation when dealing with new business models is extremely risky at this point. The likelihood of using it to find a sufficient willingness of customers to pay is very low.

3.1 The business model landscape is becoming increasingly complex

We experienced the multitude of options in a large researcher-entrepreneur consortium regarding *Industry 4.0 business models*: On behalf of the Federal Ministry for Economic Affairs and Energy, fellow researchers listed a total of 74 conceivable business models in a business model handbook concerning Industry 4.0 (see Figure 20).[5]

Based on the figure, we can see that the diversity of the business model options is simply too extensive to determine in detail whether they would all be sensible in a company's market environment.

[5] Gausemeier, Wieseke, Echterhoff, Isenberg, Koldewey, Mittag, Schneider (2017), *Corporate Success with Industry 4.0: Integrative Planning of Business Models and Value Adding Systems*, Heinz Nixdorf Institute, Paderborn, sponsored by the Federal Ministry for Economic Affairs and Energy.

Fig. 20 The complexity of business model analyses

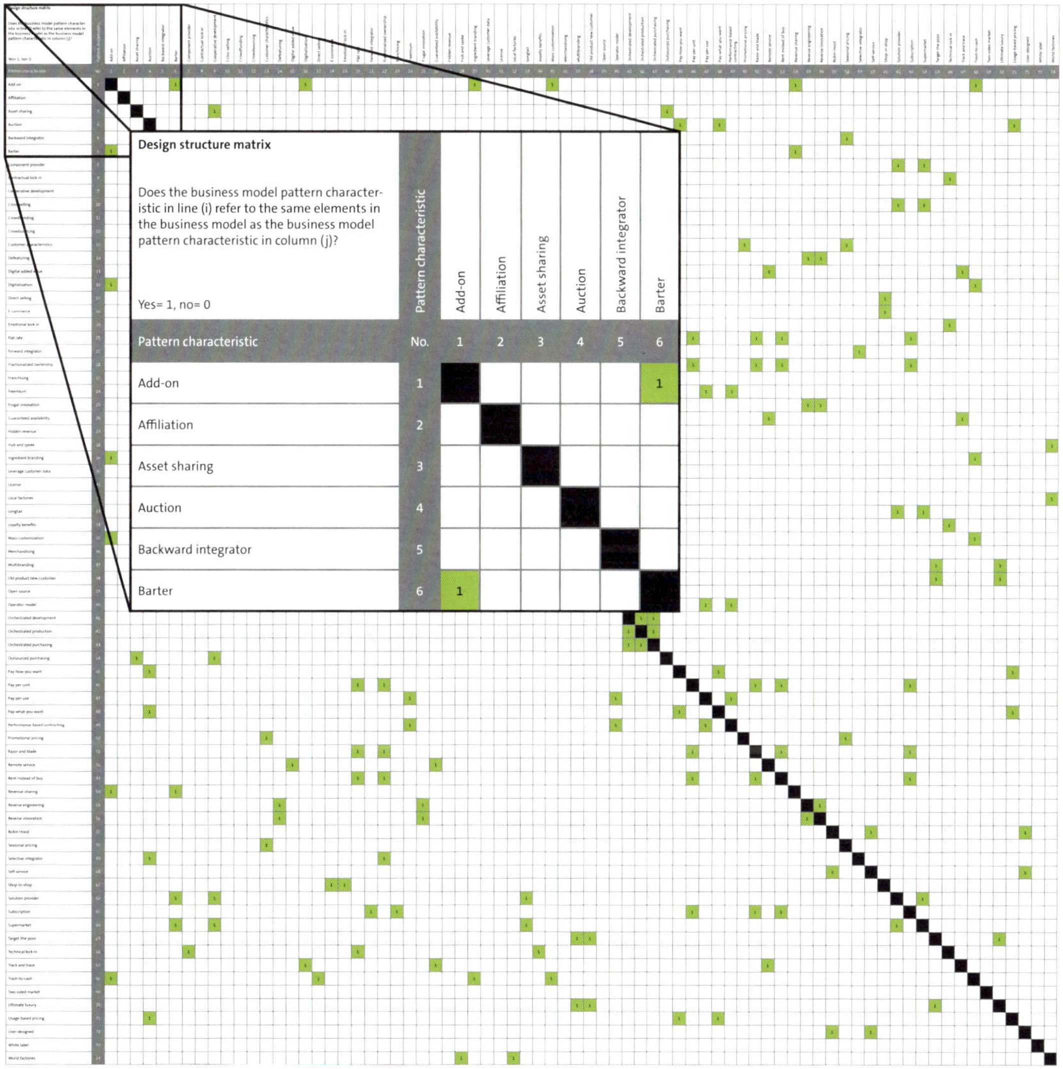

74 specific business model patterns were analyzed and assessed as regards their similarity. 2,664 individual assessments enabled the identification of **compatible patterns.**

Source: Gausemeier, Wieseke, Echterhoff, Isenberg, Koldewey, Mittag, Schneider (2017), *GEMINI – Business Models for Industry 4.0*

Information box: Digitalization as a driver of solution business models

Fig. 21 Digital elements of solution business models

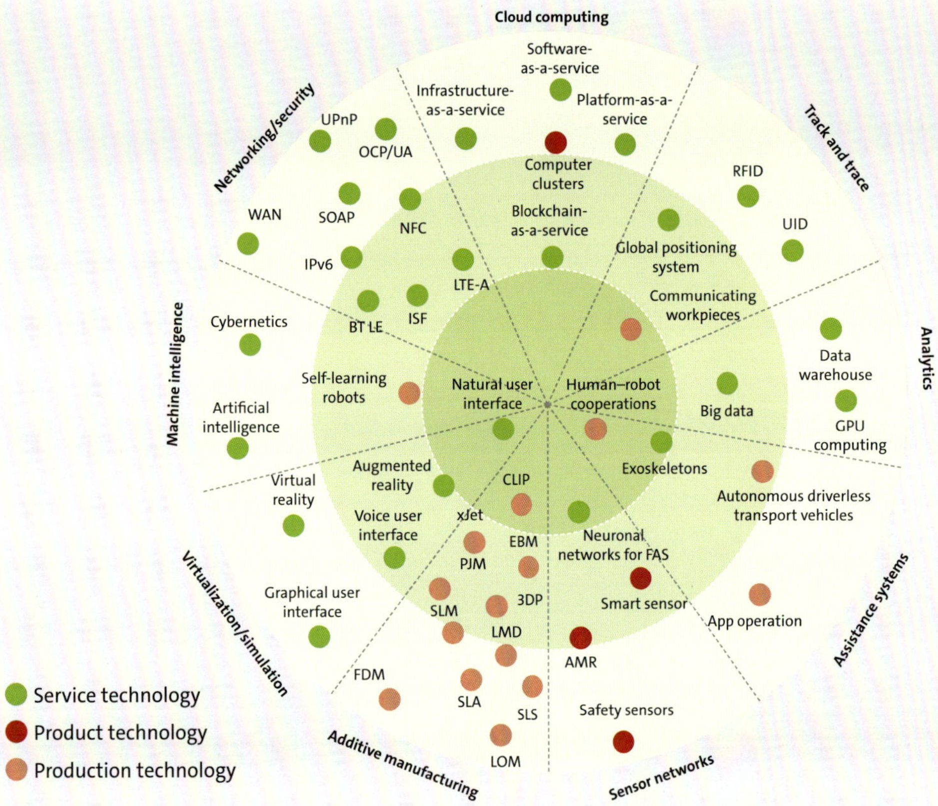

Source: Gausemeier, Wieseke, Isenberg, Koldewey, Mittag, Schneider (2017), *Corporate Success with Industry 4.0: Integrative Planning of Business Models and Value Adding Systems*, p. 19.

In Chapter 1, I explained the role of digitalization as one of the three driving forces behind Death Valley markets. It fosters market transparency and is therefore the gateway for new cost leaders.

At the same time, the possibilities offered by new business models have multiplied as a consequence of digitalization. Today, the majority of solution business models therefore develop their benefit thanks to digital components. In this regard, **digitalization can be observed in a second role**.

In this second role, digitalization helps companies to evade price and product transparency. The majority of the solution business model examples that I list in this book are based on digitalization elements. The aforementioned fellow researchers identified 48 digital tools that are used in solution business models (see Figure 21). These include software-as-a-service solutions, big and smart data analytics, RFID-supported asset management, software for operations management, and sensor-supported warning systems for predictive maintenance as well as cyber-physical production systems.

3.2 How can the complexity of the business model landscape be reduced?

It appears that we need a model that is not swamped in scientific details. And precisely this involves a major challenge because there are numerous differentiation dimensions that can be used for this purpose. The aforementioned 74 business models alone are oriented to various dimensions.

If we want to reduce the complexity of the business model options to a manageable level, we first have to answer the following question: **What actually is a business model?**

Rather than going into the numerous attempts at a scientific definition here, I want to emphasize what is essential. In order to decide on a business model, we have to answer the following questions:

1. What is marketed?
2. To whom is it marketed?
3. For which added benefit are customers willing to pay?

The third question is the most important. All a business model does is to answer the question of what customers are willing to pay for and which benefit can be used to maximize this willingness to pay. In their book *The Business Model Navigator*, Professor Gassmann and his colleagues call this the "revenue mechanism."

Or to put it differently: At which points of the value chain can we generate the biggest benefit for customers and how will they compensate us for it? This involves the search for how we can help customers to maximize the performance of their own value chain. At which points can we help them by assuming responsibility for parts of their processes?

However, the third question also involves being familiar with the other side of the coin, namely: *What are customers **not** willing to pay for?* This is an acid test, particularly for new business models, because it involves both an opportunity and a risk.

- *The opportunity* involves establishing new business models around the lack of willingness to pay. *Hilti*, for instance, recognized many customers' lack of willingness to pay in terms of owning their own equipment and invested in availability, repair, and logistics instead. To do this, *Hilti* established a successful business model that encompasses leasing the equipment, and it is marketed under the name *Hilti Fleet Management*.
- *The risk* of a lack of willingness to pay involves expensive investments in the wrong business models. This is what happens when investments are erroneously made in expensive service improvements that the majority of customers do not regard as having any major added value. This would constitute a Death Valley risk. To avoid this, the business model audit, which I introduce in Chapter 6, encompasses a *Death Valley analysis*.

Two dimensions with which we can reduce the business model options to a manageable level were determined from the variety of our market analyses. The first crucial dimension is the customers' willingness to pay (see Figure 22). The second dimension involves the scope of offerings presented to customers. I hereby mean the scope of performance that an offering encompasses.

Fig. 22 Two dimensions for structuring business models

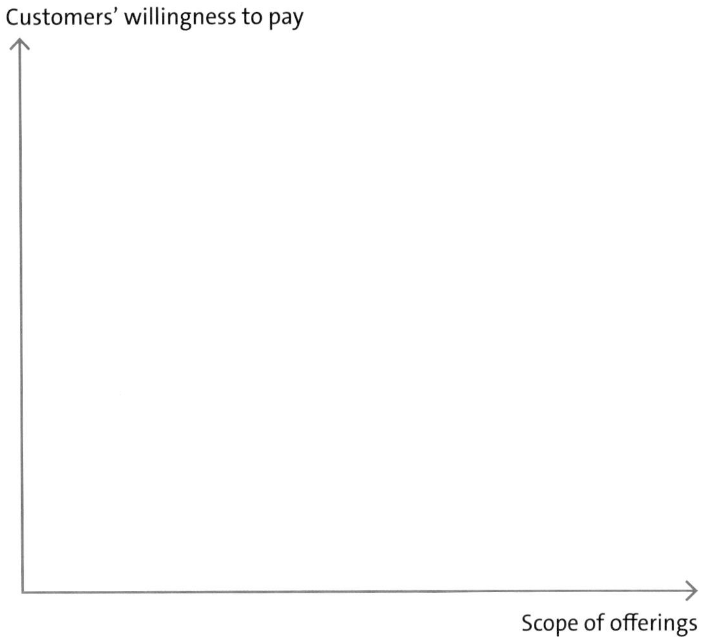

In order to group business models, we have to understand the interaction of these dimensions and know what effect changing the scope of offerings has on customers' willingness to pay.

But what form does this relationship take in reality? Let us first go through two scenarios that many managers have in their minds in this regard.

What many people believe: The myth of the easily calculated market that rewards consistent improvements in offerings. Many decision-makers believe in a linear market response to increases in performance. That would be nice and simple. Each improvement would result in a consistent increase in customers' willingness to pay (see Figure 23).

Fig. 23 The easily calculated market

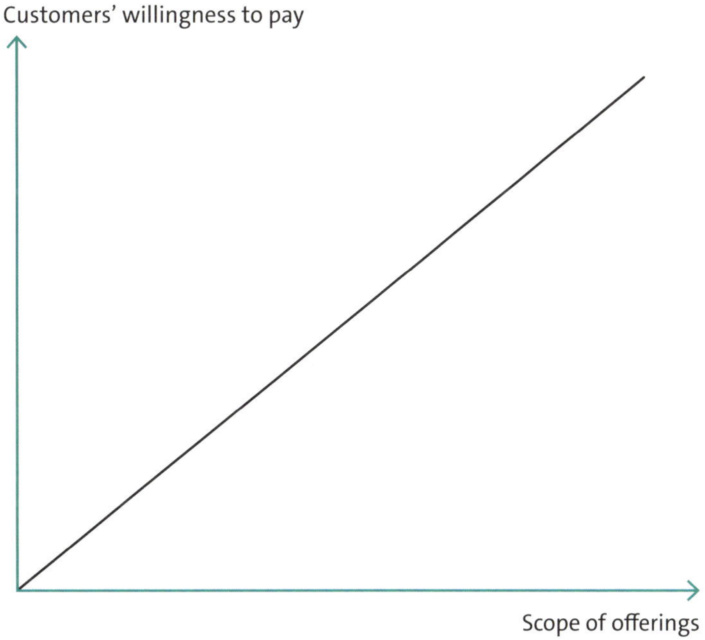

Often, this is based on a fairness logic: *"If we invest more in our offerings, the customer should also be willing to pay more."*

This thinking is particularly widespread in companies dominated by engineers. Statements such as *"We're just designing an application that will offer customers lots of additional functions"* are based on the assumption that each improvement in performance is rewarded by customers.

The belief in simple, linear relationships has a seductive appeal and is based on the assumption of honest pay for honest work and that the customers' willingness to pay is proportionate, or linear, to the scope of offerings.

But is that really the case? Are you aware of a market in which the relationship between the scope of offerings and the willingness to pay is linear, one in which each additional service consistently results in an increased willingness to pay? In all the years of researching for this book, I still haven't come across a single one.

Instead, market reality offers us *non-linear* relationships. But what form do these usually take?

What many people would like: A hyper-rewarding market that exponentially recompenses improvements in offerings. Let us first look at a non-linear relationship that could occasionally be found beforehand: The exponential relationship (see Figure 24).

An exponential relationship means that performance efforts only really pay dividends at a certain level. Before that level, the customers barely respond. After it, however, customers increasingly reward them.

This phenomenon was occasionally seen in private banking and in the luxury goods sector, for instance. Wealthy private customers barely acknowledged standardized offers but were receptive and willing to pay for individual, highly qualified solutions.

STRATEGY

Fig. 24 The hyper-rewarding market

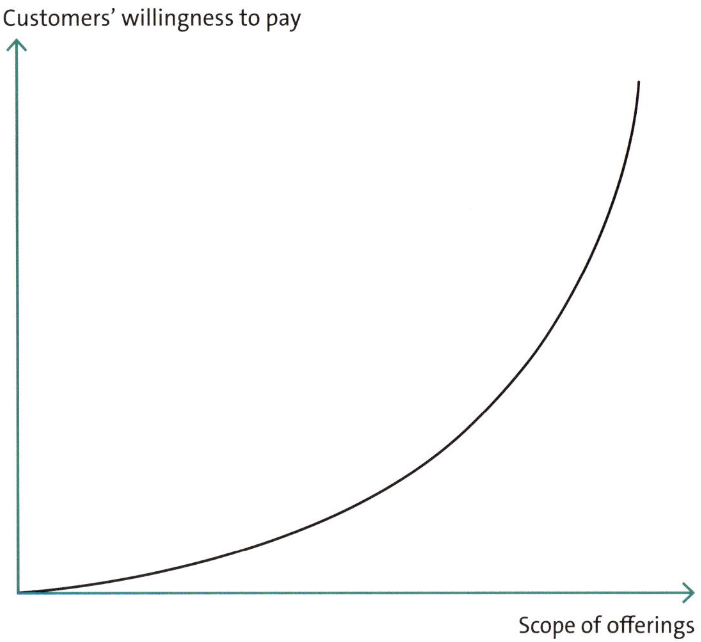

Such hyper-rewarding markets are rarely found today. Following the first wave of digitalization around the turn of the millennium, increases in service were no longer rewarded exponentially in many markets. Today, such conditions are found, if at all, in a very few niche markets.

Nevertheless, the exponential relationship remains the managers' ideal. There are also a lot of marketing consultants who fuel the belief that an exorbitant willingness to pay might arise from individual excitement factors about an offering. In reality, this aspiration leads to bad investments.

So let us turn to a form of relationship that comes much closer to today's reality.

The harsh reality: Stepped reactions to increases in offerings. Even before the new millennium, a stepped relationship had developed between the scope of offerings and the willingness to pay in many markets. Figure 25 shows an example of the basic relationship. In it, a distinction can be made between four different sections:

- There is not yet a sharp increase in customers' willingness to pay in section A. The minimum performance level that customers expect is undershot here. Every supplier should avoid this area.
- In section B, there is a steep increase in customers' willingness to pay up to the point at which customers' basic expectations are fulfilled.
- In section C, there is a plateau on which the willingness to pay stagnates and merely continues to move sideways. This area also has to be avoided.
- Section D again shows a sharp increase in customers' willingness to pay. A disproportionate investment in the improvements of offerings is worthwhile in this area. This is where companies enter the premium segment.

Do you still regard that as a boring scientific finding? The logic of the relationship curve shown here is dynamite for management theory because it contradicts many established approaches. I return to it later (see Chapter 4, p. 98).

Fig. 25 Stepped reactions to increases in offerings

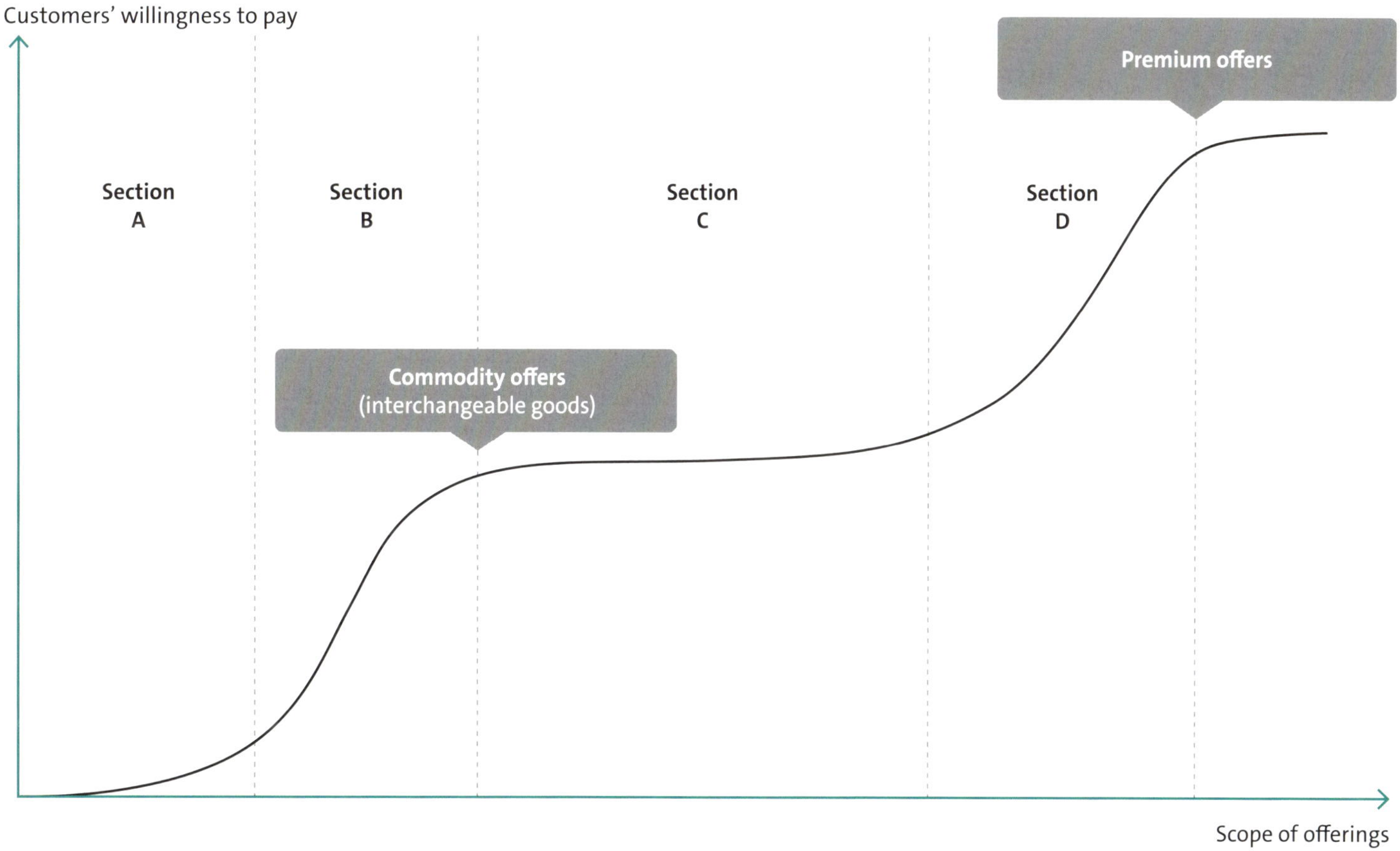

Today, more business models are being developed from this stepped basic market response (see Figure 26). The scope of offerings is continuing to increase and a gradual shift toward solution business models is taking place. The driving forces behind this are the Death Valley risks of transparent markets, which I described in Chapter 1. The route to the willingness to pay of green hill markets is being sought with solution business models.

Meanwhile, multi-step relationships, which go far beyond the commodity (interchangeable goods) vs. premium differentiation, between the scope of offerings and the willingness to pay, have developed in many markets. The more a market is developed, the more solution steps we find.

Fig. 26 The development of new business model steps

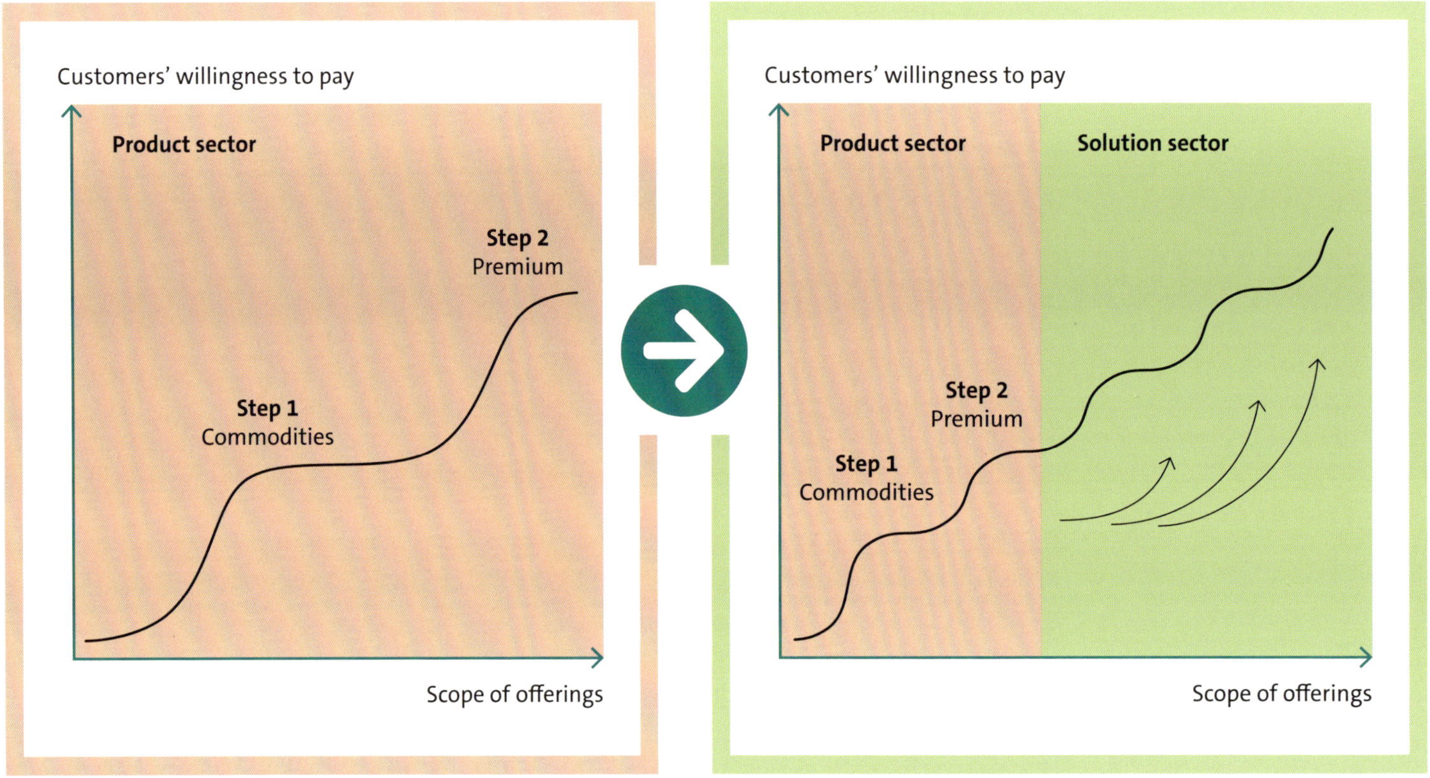

As a rule, the assumption of responsibility and the related benefit increase the further we move toward complex solution business models. This presents a range of fundamental options with which we have to deal: Are we still selling a product or are we earning from the services we provide to the customer with it? Do we lease our machines to customers and then bill them according to *outcome-based pricing*? Do we go one step further and operate our machines using our personnel on site at the customer's location? Do we take over skills that were previously in the customer's realm? Or do we go even further and link our value creation completely to that of the customer?

Information box: Development of new business model steps using the example of *Wilo*

Let's take a look at *Wilo* as an example. *Wilo* operates in the market for pumps and pump systems that are used for heating, refrigeration, air conditioning, water supply as well as wastewater disposal and cleaning.

In the commodity segment, *Wilo* offers standard pumps such as the *Wilo Yonos Pico*. These are equipped with the essential basic elements such as a manual venting function for ventilating the rotor chamber, an LC display with precise menu guidance, and pre-selectable control modes for optimal load adjustment.

Wilo still offers standard pumps in the premium segment, but these are equipped with an increased scope of offerings such as automatic reduction mode, automatic ventilation of the rotor chamber, integrated motor protection, and TÜV-tested high efficiency.

For some years now, *Wilo* has been investing extensively in solution business models. These include connective pump solutions such as the *Wilo-Assistent app* solutions. These enable the selection, administration, commissioning, and monitoring of the pumps and the accompanying services before, during, and after installation. *Wilo* has also developed a business model aimed at autonomously supplying entire water industry plants with water and controlling them remotely (see business model profile no. 10, Chapter 7).

Fig. 27 The development of new business model steps using the example of *Wilo*

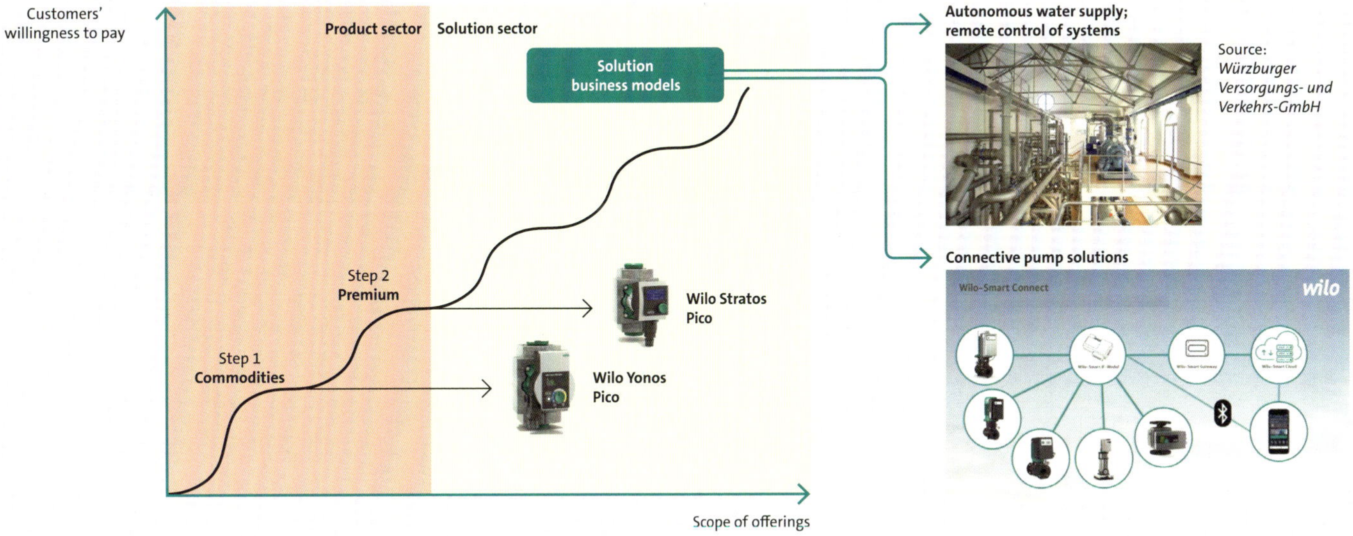

WHY SALESTEGY?

STRATEGY – FINDING THE RIGHT BUSINESS MODELS

PRACTICAL EXAMPLES

SALES – IMPLEMENTING NEW BUSINESS MODELS IN SALES

Chapter 4
The five business model steps in today's markets

Message of this chapter

Five business model steps exist in today's markets.

What should you have learned by the end of the chapter?

- The five business model steps can be designated as follows:

  Step 1: Commodity business models

  Step 2: Premium business models

  Step 3: 1st-step solutions

  Step 4: 2nd-step solutions

  Step 5: Super-complex solutions (SCS)

- This categorization is dynamite for management theory because it contradicts some of the established approaches in the literature.

In this chapter, I provide an overview of the five business model steps that can be found today in highly developed B2B markets. I also deal with the explosive power of this step logic's arguments.

DEATH VALLEY MARKETS

Strategy of minimum costs and prices

GREEN HILL MARKETS

Strategy of solution business models

FINDING

Finding solution business models

CHAPTER 3

Reducing complexity

CHAPTER 4

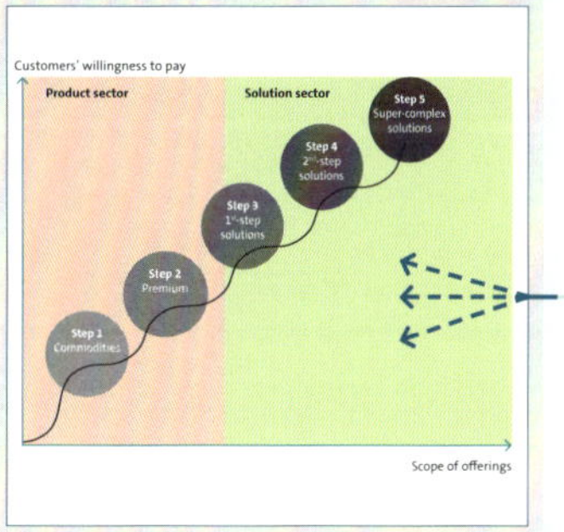

Distinguishing between five business model steps

CHAPTER 5

SNOOFY factors

Using SNOOFY factors for differentiation

CHAPTER 6

Audit: Identifying opportunities for competitive advantages in the five steps

4.1 Overview: Which business model steps are there in today's markets?

Five business model steps can currently be encountered in markets (see Figure 28). Two of them can be assigned to the classic product sector and three to solution-based business models. The *commodity* and *premium* steps are found in the product sector. In the solution sector, I distinguish between *1st-step solutions*, *2nd-step solutions*, and *super-complex solutions*.

Fig. 28 Highly developed markets with five business model steps

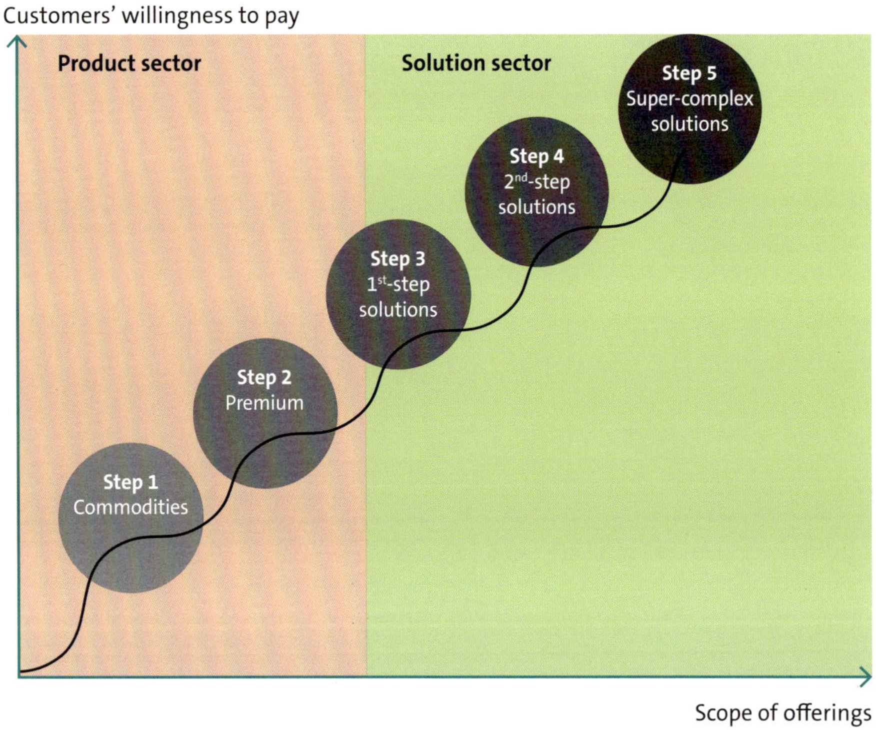

Exactly *HOW* individual business models are assigned to one of the five business model steps will be dealt with in Chapter 5. It also contains instructions for determining the steps on which your business models are located.

Let us first turn to the definition of the five business model steps.

Step 1: Commodity business models

Commodities are standard products or services with a specific, minimum level of quality. Since they are available from various suppliers in comparable quality, they are regarded as interchangeable.

Within very tight limits, commodities were originally considered to be raw materials (such as sugar, cereals, chemical and metallic raw materials) that were traded on stock exchanges. They now also include interchangeable standard products and services. Therefore, interchangeable commodities can not only be raw materials, but also basic versions of technical devices, machines, and even services.

Definition of commodity business models:

Commodities are interchangeable standard products or services that are available from various suppliers in comparable quality.

Due to this standardization, the majority of customers consider the price to be crucial. Additional elements do not play any role.

Step 2: Premium business models

Premium products are a well-known business model step in high-wage industrialized nations. The competitive advantage compared to commodities is continually being sought in increasingly new quality aspects that the competition does not yet offer. Attributes such as quality, functionality, ergonomics, longevity, and design are continuously optimized. Premium business models are the result of strong research and development departments and strong brands that symbolize quality leadership.

The success mechanisms of the premium business models are the same as they were 25 years ago and have been dealt with exhaustively in relevant strategy books. In our context, they are immaterial.

Definition of premium business models:

With premium products or services, suppliers seek to find their competitive advantage over commodities in increasingly new quality aspects that the competition does not yet offer.

Solution business model steps. The most extensive market developments are taking place in the solution sector. Additional business model steps are continually evolving here.

If we take a look at the market with the three solution business model steps, it initially appears complex. At first glance, it may even seem unrealistic that such different steps exist. However, that certainly isn't the case.

The 18 business model profiles in Chapter 7 show examples of companies from various industries for each of the three solution steps. They underpin just how differentiated the business models now are.

Step 3: 1st-step solutions

I refer to situations in which initial additional services are offered as solutions around the core product as 1st-step solutions. 1st-step solutions are characterized by the fact that they facilitate customer process steps in the value chain (through after-sales services, training courses, or maintenance contracts, for instance). They are therefore concerned with enhancing effectiveness and efficiency in the customers' value chains.

Taking over minor process steps is also conceivable in this process. However, focus is placed on the supplier's *support* function for the customers' value chains. The scope of the affected customer value chain processes is still comparatively small.

Definition of 1st-step solutions:

In 1st-step solutions, suppliers offer a benefit by making minor process steps in the value chain easier for their customers.

Step 4: 2nd-step solutions

I refer to solutions that take a further, major step toward the profitability of customer processes as 2nd-step solutions. Here, the suppliers intervene more extensively into the customer processes. In contrast to the 1st-step solutions, they not only help the customers in their processes by means of minor support functions, but now they also intervene in neuralgic processes in order to optimize the customers' value chains. To accomplish this, suppliers usually take over process steps at the customer's site (such as warehouse management) themselves in the case of 2nd-step solutions. As the practical examples in Chapter 7 show, such steps can be taken over at numerous points.

Responsibility for the results of the processes is also increasingly assumed. This necessitates new pricing models. *Outcome-based pricing models*, in which the price is based on the result of the solution, are increasingly being employed for this. Often, results are actually guaranteed to customers in 2nd-step solutions.

Definition of 2nd-step solutions:

In 2nd-step solutions, suppliers intervene even more comprehensively into their customers' value chains and actively assume responsibility both for extensive process steps and for the results of the processes.

Step 5: Super-complex solutions (SCS)

Super-complex solutions can be regarded as the final step of the solution evolution. Their scope of performance is so extensive that the customer transfers significant processes to the supplier. This can even include taking over central processes and production plants (such as software for managing diagnostic processes for patients in a hospital). Classically, super-complex solutions involve maximum integration of the supplier and customer companies. This means that the supplier and customer network personnel, machines, and IT systems.

Super-complex solutions are focused on the primary processes of the customer value chain, such as inbound logistics, operations/production, marketing and sales, outbound logistics, and customer service. However, more minor supporting activities also continue to play a role.

The objective of super-complex solutions is a maximum increase in the customer benefit. Incidentally, my colleague Professor Olaf Plötner calls the most complex step of the solution business models "complex service solutions."[6] The content classification is consistent with that used here. Since we know one another well, both terms originate from the same conceptual realm.

Definition of super-complex solutions:

Super-complex solutions are the final step of the solution evolution. In them, suppliers extend their range of services to such an extent that customers' core processes are shifted to them and/or the supplier and customer are extensively integrated. The objective is maximum customer benefit.

STRATEGY

[6] Plötner (2012), *Counter Strategies in Global Markets*, Palgrave Macmillan, pp. 89 ff.

Super-complex solutions are increasingly reaching new heights. In the Industry 4.0 vision of networked factories that self-evolve using artificial intelligence, super-complex solutions are taken to the extreme. The boundaries between suppliers and customers disappear entirely in such integrated value-creation processes.

Information box: On which step are your business models located?

The following work template is useful for an initial classification of your business models into the five business model steps.

Please mark the business model steps in which you are currently active or are likely to be active in three years with a checkmark in the respective column.

	Today	In three years (planned)
Commodities (interchangeable standard products, e.g. simple pipes and C-parts)	☐	☐
Premium products (high-quality products, e.g. high-quality air conditioners)	☐	☐
1st-step solutions (facilitation of customers' minor process steps, e.g. through after-sales services, training courses, or maintenance contracts)	☐	☐
2nd-step solutions (takeover of extensive processes such as warehouse management for customers)	☐	☐
Super-complex solutions (takeover of core processes for customers, e.g. software for managing diagnostic processes for patients in a hospital)	☐	☐

How are your total sales distributed over the aforementioned business model types today? Please refer only to the business model types that have a checkmark under "Today" in this table.	
The percentage values indicated should total 100.	
Commodities (interchangeable standard products, e.g. simple pipes and C-parts)	________ %
Premium products (high-quality products, e.g. high-quality air conditioners)	________ %
1st-step solutions (facilitation of customers' minor process steps, e.g. through after-sales services, training courses, or maintenance contracts)	________ %
2nd-step solutions (takeover of extensive processes such as warehouse management for customers)	________ %
Super-complex solutions (takeover of core processes for customers, e.g. software for managing diagnostic processes for patients in a hospital)	________ %
	100%

How will your estimated total sales *in three years* be distributed over the aforementioned business model types? Please refer only to the business model types that have a checkmark under "In three years (planned)" in this table.	
The percentage values indicated should total 100.	
Commodities (interchangeable standard products, e.g. simple pipes and C-parts)	________ %
Premium products (high-quality products, e.g. high-quality air conditioners)	________ %
1st-step solutions (facilitation of customers' minor process steps, e.g. through after-sales services, training courses, or maintenance contracts)	________ %
2nd-step solutions (takeover of extensive processes such as warehouse management for customers)	________ %
Super-complex solutions (takeover of core processes for customers, e.g. software for managing diagnostic processes for patients in a hospital)	________ %
	100%

Now that you have been introduced to the five business model steps, you may also be asking yourself one of the questions that managers often put to us in this context:

"How do we get to the next business model step?" To answer this question, it is necessary to know the six "SNOOFY" factors, which I deal with in the following chapter.

"So are higher business model steps always better than lower ones?" Often, we are also asked whether it is not better per se to forge ahead to the highest possible business model step as quickly as possible. After all, the "higher" business model steps offer protection against commoditization and price erosion thanks to their complexity. Nevertheless, I advise against this. Instead of skipping business model steps (going directly from premium to 2nd-step solutions, for example), I recommend a step-by-step procedure. Evolution from one business model step to the next involves a significant effort and risks. I discuss this in Chapter 6 (p. 148), and we will see that each business model step entails its own Death Valley risks.

4.2 Statistics on the distribution of the business model steps

Our business model study involving 845 B2B companies provided findings on the distribution of the business models over the business model steps. At this point, I would like to report on four of these findings.

The first finding is that more than half of the analyzed companies operate simultaneously on two or more business model steps. Around 20% actually operate on three or more steps with their business models.

The second finding concerns the distribution of today's business models compared to those that the analyzed companies are planning in three years. This is shown in the following figure. It reveals a clear trend toward the solution business model steps. At present, numerous companies are planning to enter the 2nd-step (+11.3%) and super-complex solutions (+13.2%).

Fig. 29 Distribution of active business models

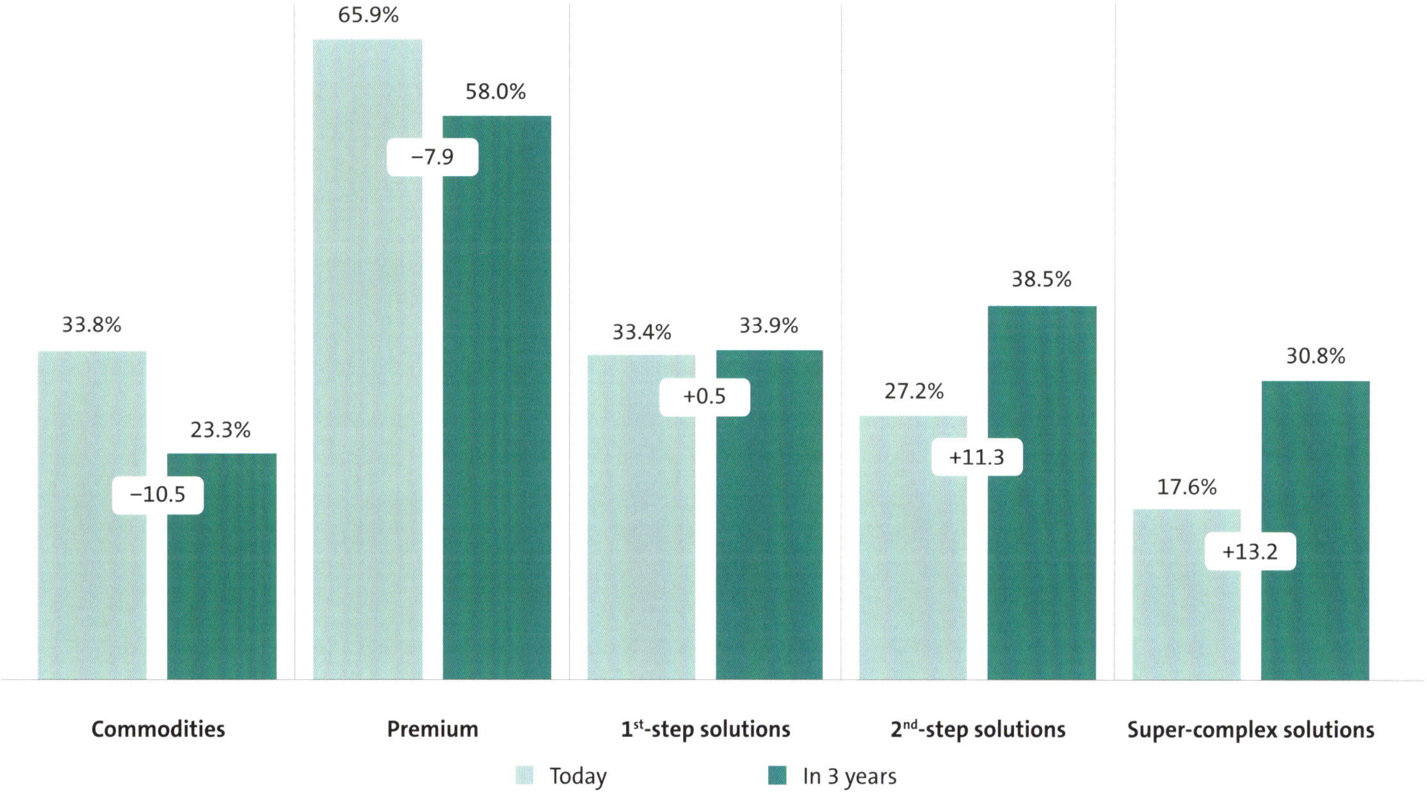

Note: As many companies operate simultaneously on several business model steps, the values add up to more than 100%.

The third finding concerns the development of sales distribution over the business model steps. As the pure existence of a business model does not yet have any bearing on its commercial importance, we also analyzed sales distribution in our study.

Figure 30 shows that the relevance of the product business models in B2B markets will decline comparatively over the coming years – measured against their sales – whereas that of solution business models will increase. However, the magnitude of the change indicates more of a slower evolution rather than an overnight market revolution on the whole.

Fig. 30 Distribution of sales between the business model steps

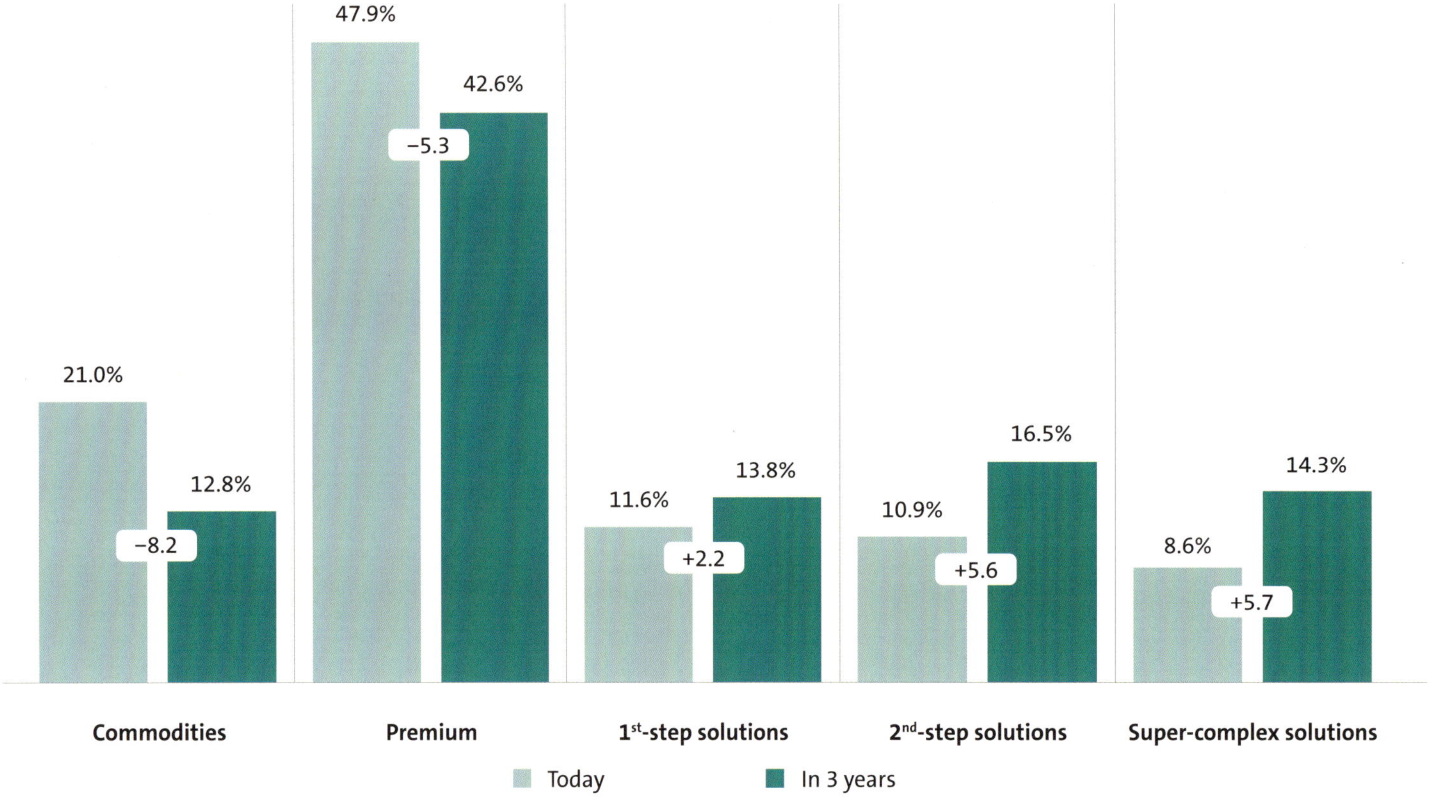

The fourth finding that our data provides are the detailed values for development in individual B2B industries. As an example, I have picked out development in the industrial goods and mechanical engineering industry here (see Figure 31). Development here is far higher than the average.

In the practical example section (see Chapter 7), I present six business model profiles from the industrial goods and mechanical engineering industry.

Fig. 31 Industry analysis on the development of business models

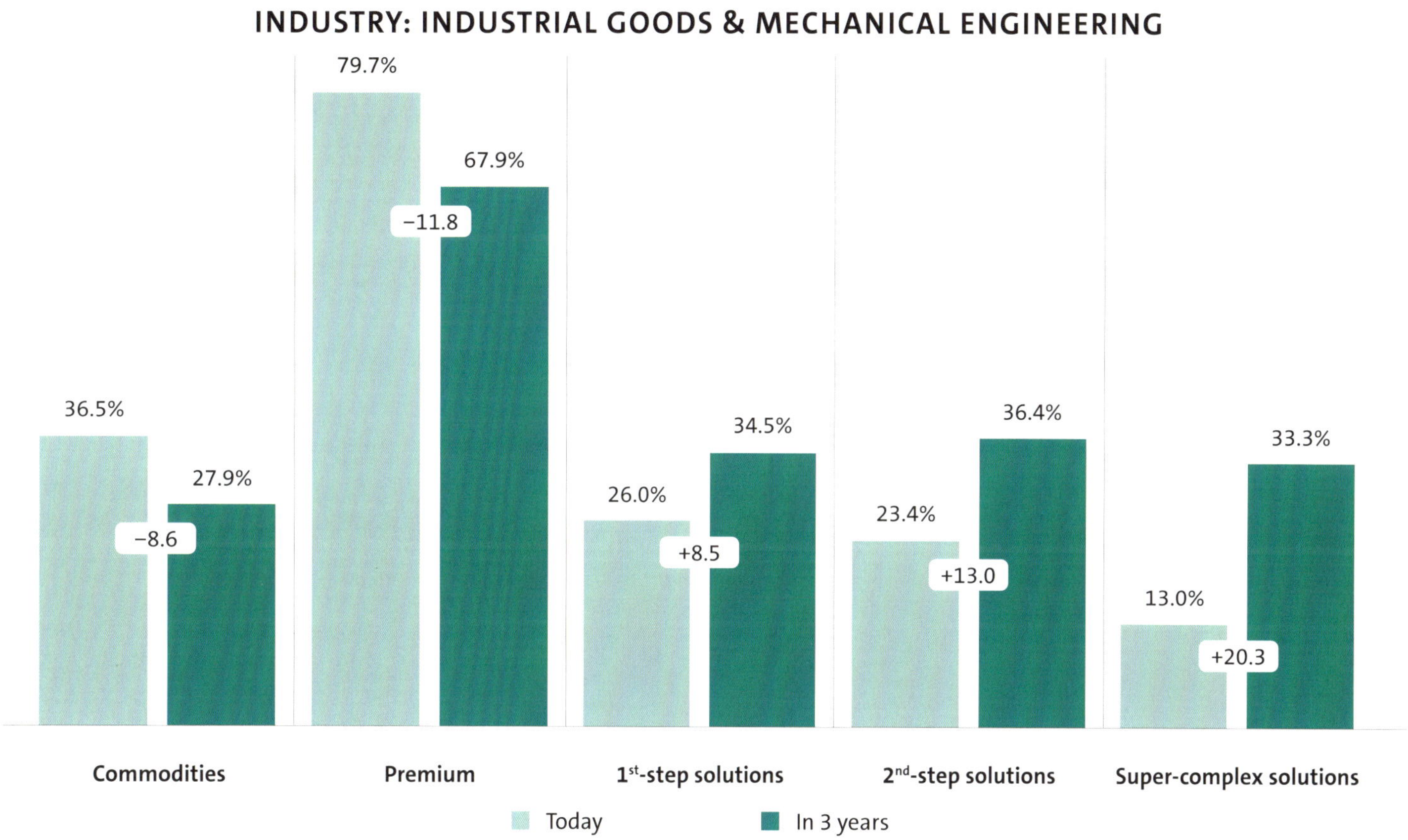

Note: As many companies operate simultaneously on several business model steps, the values add up to more than 100%.

4.3 Dynamite for management theory: Revolution or evolution of business models?

The explosive power of the step logic is due to the fact that it contradicts current mainstream beliefs. At the numerous management congresses at which I am a guest, keynote speakers who refer to themselves as "evangelists of digitalization" talk with something akin to religious fervor about the digital REVOLUTION of the markets.

At this point, I would like to clearly and unmistakably counter this high-profile assertion that the markets are being overwhelmed by a digital REVOLUTION. Of course, the term REVOLUTION sounds spectacular and ensures large audiences at management congresses. However, I am increasingly shocked at the panic that such statements trigger among many companies. Panic is the very last thing that walkers need when navigating through Death Valley.

In the majority of B2B markets, I am observing an EVOLUTION. While that sounds less spectacular, this gradual process involves an insidious risk that is all the more treacherous because it often crops up under the management's radar. I therefore state that the majority of companies do not fail because of market REVOLUTIONS but because of the EVOLUTION of their business models in the transition to the next business model step. I deal explicitly with this in Chapter 6 (step 4) and Chapter 8.

It also explains the difference between my approach and the "Blue Ocean" approach (see information box). The Blue Ocean approach sees commercial success in the creation of completely new markets. My approach sees commercial success in the step-by-step development of business models within markets.

Of course, the B2C sector also includes markets such as the cellphone market, in which companies such as *Nokia* are forced out within a few years, and it is possible to talk about a REVOLUTION in the negative sense. But developments in the majority of markets, particularly in B2B markets, are still considerably more protracted. Do not misunderstand me: An EVOLUTION can also completely change markets. It simply takes far longer. And it is taking place in B2B markets within the five business model steps that I have outlined.

Hand on heart: Is a digital **re**volution taking place in your market? Will your market change completely in less than five years? Or is it more of an **e**volution, in which this development will progress more slowly?

Information box: What is the difference compared to the "Blue Ocean Strategy" concept?

Readers who have dealt with the groundbreaking "Blue Ocean Strategy" concept may possibly be wondering whether solution business models and the green hill analogy are not comparable to the *Blue Ocean business models*. With sales of more than four million copies, the Blue Ocean Strategy book is one of the best-selling strategy books of the past two decades.[7]

The feature common to both approaches is the search for innovative business models. The major difference is in ***how*** the innovations are achieved. Blue Ocean business models involve *market revolutions* and function according to entirely novel ground rules. Conversely, solution business models are based on *market evolution* and look for innovations through additional step-by-step developments.

The background to my green hill analogy lies in the observation that revolutionary business models are only implemented very rarely compared to evolutionary business models. This observation particularly applies to the B2B markets analyzed in this book.

	Blue Ocean business models	Green hill business models
Form of change	Revolution → radical break	Evolution → gradual steps
Market focus	Creation of new markets	Evolution in the same market
Application area	B2C > B2B	B2B > B2C
Relevance of SALESTEGY	Not relevant per se	Relevant per se
Status of marketing and sales	Marketing > sales	Sales > marketing (integration necessary)

[7] Kim, Mauborgne (2015), *Blue Ocean Strategy, Expanded Edition: How to Create Uncontested Market Space and Make the Competition Irrelevant*. Harvard Business Review Press.

WHY SALESTEGY?

STRATEGY – FINDING THE RIGHT BUSINESS MODELS

PRACTICAL EXAMPLES

SALES – IMPLEMENTING NEW BUSINESS MODELS IN SALES

Chapter 5
How do you get to the next business model step?

The six "SNOOFY" factors are crucial

- **S** – Servitization
- **N** – Nurture customer processes
- **O** – Operate processes for customers
- **O** – Outcome-based pricing
- **F** – Far-reaching relationship
- **Y** – Yield customer value-in-use

The SNOOFY business model test

Message of this chapter

The six "SNOOFY" factors determine the step on which a solution business model is located: 1st-step solution, 2nd-step solution, or super-complex solution.

What should you have learned by the end of the chapter?

- The "SNOOFY" factors enable business models to be assigned to the solution business model steps (1st-step solutions, 2nd-step solutions, and super-complex solutions).
- "SNOOFY" =
  Servitization
  Nurture customer processes
  Operate processes for customers
  Outcome-based pricing
  Far-reaching relationship
  Yield customer value-in-use
- The *SNOOFY business model test* can be used to determine the SNOOFY score per business model; this then acts as an indicator for the business model step.
- In turn, the business model step impacts the strategy and sales:
  - **Strategy:** Decisions regarding the strategic positioning of one's own business models (see Chapter 6)
  - **Sales:** Decisions regarding the entire spectrum of sales management (see Chapters 9–14)

DEATH VALLEY MARKETS

Strategy of minimum costs and prices

GREEN HILL MARKETS

Strategy of solution business models

FINDING

Finding solution business models

CHAPTER 3

Reducing complexity

CHAPTER 4

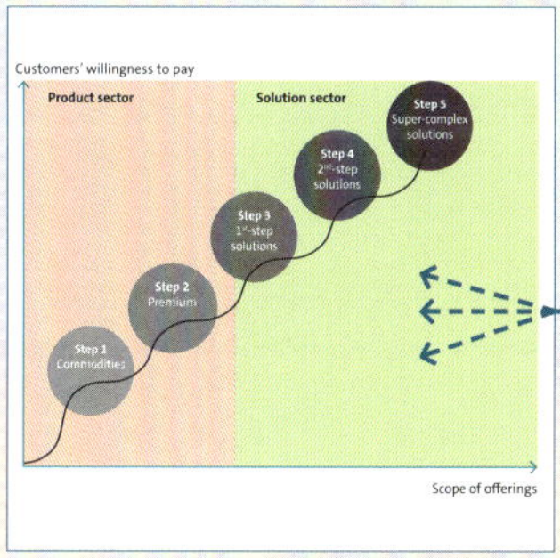

Distinguishing between five business model steps

CHAPTER 5

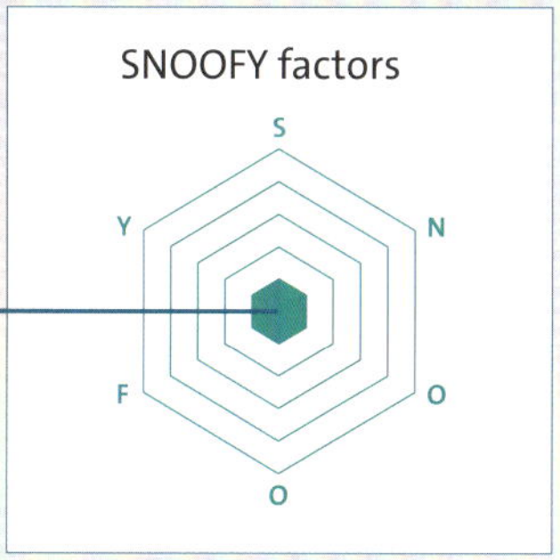

Using SNOOFY factors for differentiation

CHAPTER 6

Audit: Identifying opportunities for competitive advantages in the five steps

Are you wondering why it is important to know which step each of your business models is located on? Because this knowledge leads to a chain of direction decisions for your organization. This applies, firstly, to strategic decisions concerning the positioning of your own business models. I discuss this in Chapter 6. Secondly, it applies to sales decisions, such as the decision whether you should use field salesforce employees and, if so, how many.

Our research work determined six factors that can be used to differentiate between business model steps in a scoring model. They help to assign your business models to the appropriate business model step.

The six differentiation factors can be abbreviated to **"SNOOFY"** and stand for (see Figure 32):

- **S**ervitization
- **N**urture customer processes
- **O**perate processes for customers
- **O**utcome-based pricing
- **F**ar-reaching relationship
- **Y**ield customer value-in-use

Fig. 32 The SNOOFY factors

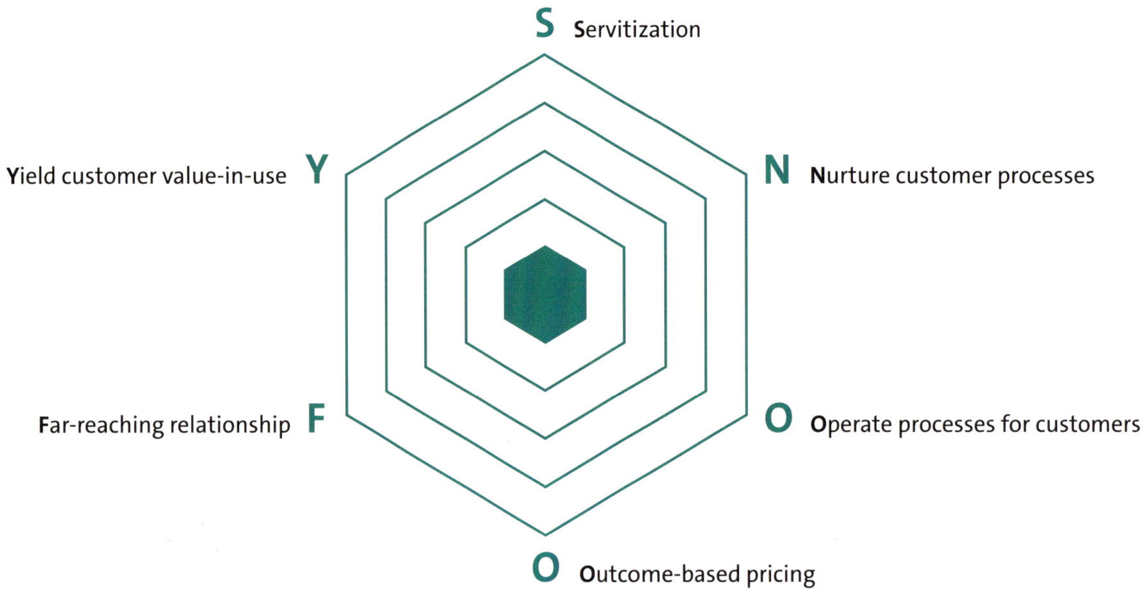

As Figure 33 shows, the values for each of the six factors increase from the commodity business model up to super-complex solutions. For the first two factors *(**S**ervitization; **N**urture customer processes)*, this progression undergoes the biggest leap in the step from product business models to 1st-step solutions. The biggest leap for the other four factors occurs in the shift from 1st-step to 2nd-step solutions. For example, our observations found that little ***O**utcome-based pricing* takes place in 1st-step solutions. However, it is usually part of the business model in 2nd-step solutions.

Hierarchy of and links between the SNOOFY factors. There is no hierarchy in the importance of the six factors, but there are logical links between the factors. For example, ***O**utcome-based pricing models* guarantee the value contribution of solutions *(**Y**ield customer value-in-use)*. Value-in-use means that the supplier optimizes customer processes by supporting or taking them over *(**N**urture customer processes; **O**perate processes for customers)*.

Fig. 33 The development of the SNOOFY factors over the business model steps (average)

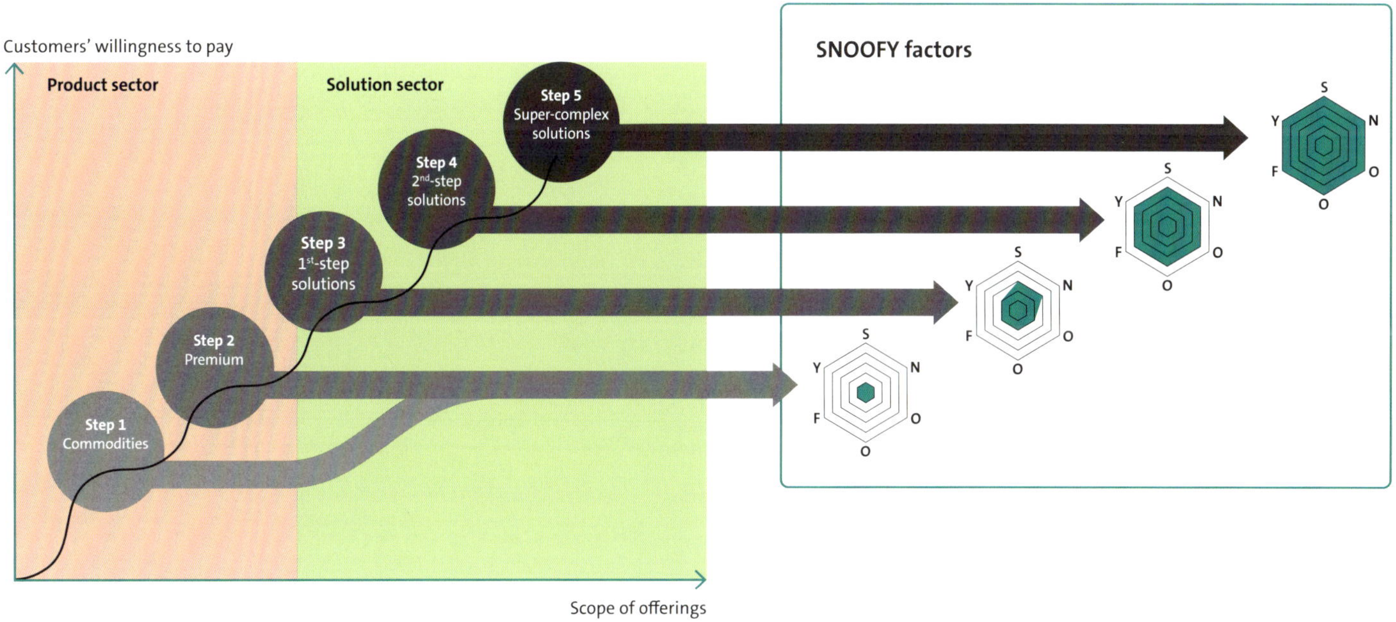

S – Servitization

The first differentiation factor in the business model steps is the service content. The service content increases as a supplier progresses from products to more complex solutions.

The service content was the first point of reference of the research into solution business models. The major research streams regarding "service-dominant logic" and "service infusion" found their starting point here. The established term "hybrid offerings" also indicates that offers consisting of product and service contents are intermingling.

Using the example of the industrial goods and mechanical engineering industry, Figure 34 shows how the service content continues to increase over the business model steps. It also shows typical business model patterns such as after-sales service and fleet management, which can be assigned to the individual steps.

The transformation from product-centric suppliers, where only service elements are added, to solution providers, where the tangible product is shifted to the background, is particularly reflected in the industrial goods and mechanical engineering industry. This is shown in the path from after-sales services to Industry 4.0.

Fig. 34 Service content along the business model steps (example of industrial goods and mechanical engineering)

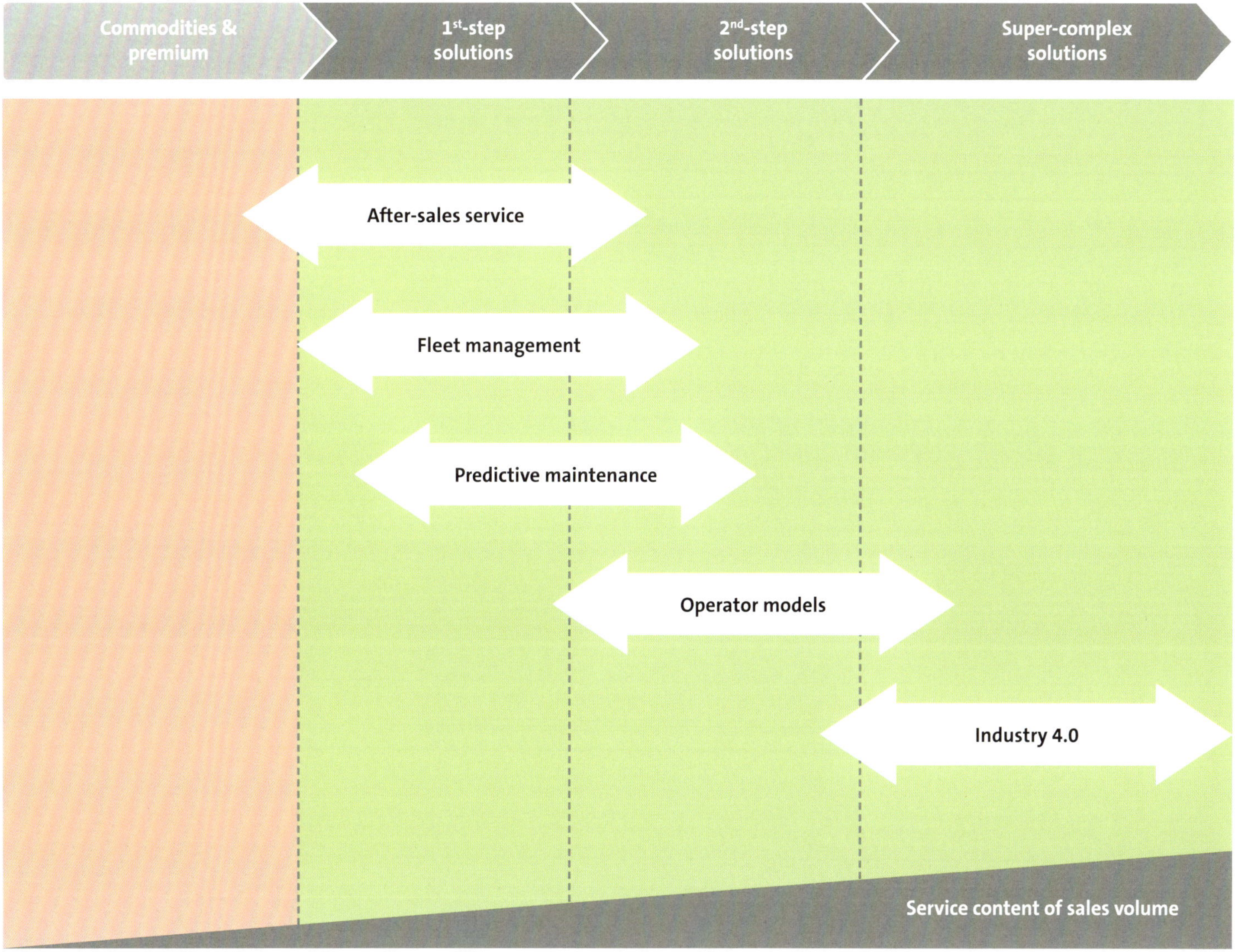

- *After-sales service.* Suppliers often start out with simple services such as after-sales services, which are offered as additions to the core product. This encompasses simple services such as maintenance, procurement of replacement parts, and repairs. Examples of this can be found in business model profiles nos. 1, 5, and 7 in Chapter 7.

- *Fleet management.* The fleet management service involves keeping a large number of components (*Michelin Solutions*, for example; information box, p. 122), assets and machines (*Hilti Fleet Management*, for example), or cars/trucks (*Mercedes PRO connect*, for example) optimally operational. An example of this can be found in business model profile no. 5, Chapter 7.

- *Predictive maintenance.* Predictive maintenance is aimed at optimizing the use of machines, systems, and vehicles by minimizing their malfunctions and downtimes. To do this, large quantities of data points are analyzed to predict failure risks. Sensor data is especially important for this. Examples of this can be found in business model profiles nos. 1, 8, and 10, Chapter 7.

- *Operator models.* In operator models, suppliers take on responsibility for the entire project management, planning, construction, financing, and operation cycle of machines, systems, buildings, and IT systems, etc. In mechanical engineering, this can mean the operation of entire automotive production lines. In the IT sector, responsibility is taken over for hosting services, virtual private networks (VPN), voice over IP (VoIP), Internet access, data backups, network security, and technical support among others. Examples of this can be found in business model profiles nos. 9, 12, and 14, Chapter 7.

- *Industry 4.0.* The aim of Industry 4.0 is maximum networking of industrial production. Comprehensive IT services are leading to intelligent and digitally networked production systems. These "cyber-physical production systems" are intended to enable the most autonomous production possible. To do this, human, machine, component, system, and logistics data is networked. Examples of this can be found in business model profiles nos. 16–18, Chapter 7.

Information box: *Voith* as an example of increasing service content

One company that has achieved a radical transformation from a product to a service supplier is the *Voith Group*. Besides many other industrial goods, *Voith* manufactures paper machines. In the past, *Voith* was a pure supplier of paper machine products in this business field. Today, the service content of its sales is already over 50%.

Personally provided, automated, and data-driven services. The service content of the business models can be subdivided into personally provided services, automated services, and data-driven services. Personally provided services would include repair services performed by technicians. Automated services include robotics services used for the repair. One data-driven service would be predictive maintenance software that avoids repair work by pointing out wear early on through data analyses.

Essentially, complex business models shifting toward 2nd-step solutions and super-complex solutions are now based primarily on automated services and data-driven services. Among other aspects, the distinction between personally provided services, automated services, and data-driven services is important when entering new national markets with solution business models (in this regard, see Chapter 13, pp. 322 ff.), as statutory regulations concerning data processing are more restrictive in some national markets than in others.

Special case of software-as-a-service (SaaS). For many classic product suppliers, entering the SaaS business marks a serious step in the direction of services. SaaS business models originate from the domain of cloud computing. Here, suppliers act as IT service providers that operate a software system or even the entire IT infrastructure for their customers.

Many production companies seek to enter the SaaS business by means of easily operated apps. *Hilti*, for instance, offers its customers an uncomplicated asset management option with its *ON!Track app*. We are familiar with highly developed SaaS solutions from major software companies such as *Salesforce*, *Oracle*, and *SAP*.

It is interesting to observe how traditional product suppliers evolve from 100% *tangible* product suppliers to 100% *intangible*, digital service providers. The transformation from a hardware to a SaaS supplier marks a major milestone for many companies.

Jeff Immelt, the former CEO of *General Electric*, summed up this trend as follows: *"Every industrial company must be a software company."*[8]

Suppliers in classic service markets are also differentiated by their automated service and data-driven service content. The total service content, which is per se 100%, does not change here, but the personally provided service, automated service, and data-driven service content varies. The surveys conducted by the service association *KVD* provide interesting findings in this regard. According to these, the percentage of companies that do not generate sales with data-based services has fallen from 46% to 21% over the past five years.[9]

---

[8] Plötner (2016), *Cost traps in business models 4.0*, ESMT Berlin, p. 8.

[9] https://www.service-verband.de/presse/data-driven-services-die-wichtigsten-trends-und-zahlen-aus-dem-service, accessed on 11.17.2020.

N – Nurture customer processes

The word "nurture" is also synonymous with "foster" and "support." ***N**urture customer processes* means fostering and supporting customer processes. The farther that providers proceed along the path to becoming solution providers, the more they support their customers in their processes.

One essential driver of this evolution involves a change of perspective on the part of the suppliers. There is a shifting away from offering a product – in which it is left up to the customer to decide the manner in which the product is to be used to generate value – toward becoming a process facilitator that enables optimally profitable process flows for the customer.

This change of perspective inevitably affects the supplier's departments and functions. Different structures, processes, and know-how are needed.

The first question to be answered is **where** processes can be facilitated for the customers. To do so, let us visualize the elements of the value chain.

Let us first deal with the classic value chain according to Michael Porter. Two starting points for solution offers arise here (see Figure 35). Firstly, they can be based on the ***primary activities*** of the customer value chain. These include inbound logistics, operations/production, marketing and sales, outbound logistics, and service. Secondly, they can improve ***supporting activities*** of the customer value chain such as the firm infrastructure, human resource management, technology development, and procurement.

Fig. 35 Starting points for solution business models in the classic value chain

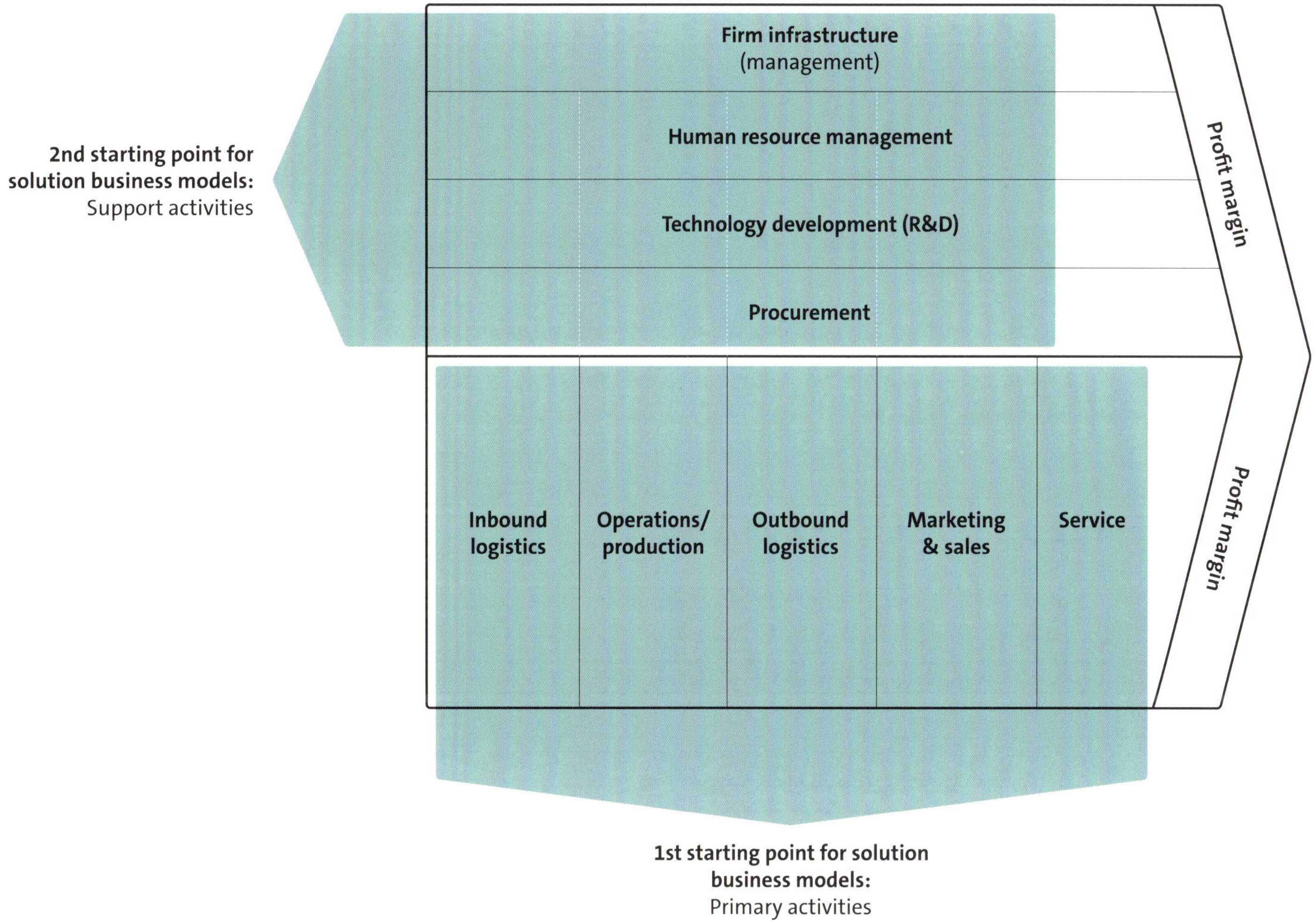

Source:
Porter (1985), *Competitive Advantage – Creating and Sustaining Superior Performance*, Macmillan, p. 37.

First starting point for solution business models: Help in improving **primary activities** in the customer value chain.

To support customers in their primary processes, many suppliers offer their customers *simple consulting services*. These are aimed at helping to optimize the profitability of the customer processes wherever the customers themselves do not have the know-how.

Today, such offers can be found for all primary processes. Many production companies start out in the solution business with *customer service*. This helps the customers through user training courses and advice on optimal maintenance of their products and machines. The nurture offers are applied primarily in the area of the customers' *operations/production*. In this regard, let us take another look at *Voith's* paper machine division. Use of the *Voith* machines is one of the core tasks of printing companies. Nevertheless, *Voith's* know-how in using the machines will always surpass theirs. By providing advice on the optimized use of the paper machines, *Voith* can therefore increase the long-term benefit to its customers. "Building Information Modeling" (BIM) solutions offer even more extensive operations/production support in the construction industry. These BIM software solutions enable suppliers such as *Hilti* and *PERI* to advise clients when planning complex buildings and can facilitate the subsequent construction process (see business model profiles nos. 6 and 11, Chapter 7).

In markets with pronounced dealer structures, some companies use support services to increase end-user demand. One good example of this is *ARDEX*, a manufacturer of chemical construction materials and the global market leader in wall surfaces, floor coverings, and parquet adhesives. In an indirect sales model, *ARDEX* markets its products to trade businesses via specialist retailers. To boost user demand for the products, the *ARDEXacademy* provides training courses for more than 20,000 tradesmen per year. *Vaillant* takes a similar approach for heating, ventilation, and air conditioning (HVAC) contractors.

Second starting point for solution business models: Help in improving the **support activities** in the customer value chain.

1st-step solutions, in particular, are often aimed at improving the support activities in the customer value chain. We are seeing an explosion of such offers in B2B markets.

One major company that offers solutions for support activities is *McKesson*. Besides delivery services, which are barely profitable any longer, *McKesson* is increasingly offering operations management consulting services for self-employed pharmacists in the fiercely contested pharmaceutical sales market. Of course, many pharmacists undergo training primarily in pharmaceutical knowledge, but not in the commercial management of a drugstore as a company. This is where *McKesson* comes into play by offering management support. Marianne Timm, Country Manager for Denmark at *McKesson*, describes this drugstore management assistance as follows:

"We divide our customers into two types – let me illustrate it with an example. Take pelicans: There are those who simply want to have their fish provided and those who want to learn how to fish. For those who want to learn, we offer our expertise and consult them on how to be more effective in their business."

Nurture solutions based on changed value chains. The options for facilitating customer processes change with the transformation of value chains in digitalized markets.

We have conducted intensive research into what form these value chain changes take and what this means for company decision-makers. To do this, the collaborative research center "TR29" was set up at the *Ruhr University of Bochum* and the *Technical University of Berlin* and was provided with more than $11.7 million in funding by the German Research Foundation (DFG). Just how cross-functional the impact of solution business models is for suppliers is reflected in the interdisciplinary makeup of the participating scientists. They came from the specialist fields of product and service engineering, human-machine systems, industrial information technology, mechanical engineering informatics, production management, machine tools and factory operations, psychology, occupational science, marketing, sales, knowledge management, automation, controlling, employee skills and leadership actions, and from quality science. The solution provider consortium consisted of 20 corporate partners such as *Trumpf, Claas, Ericsson AB, Airbus Operations GmbH, TATA Interactive Systems GmbH,* and *Carat Robotic Innovation GmbH.*

Our department's task involved investigating the changes in business models in today's industrial landscapes and in determining the changes necessary in sales. Figure 36 shows such a value chain.

As can be seen, the process flows become more complex on the whole compared to the classic value chain, and the transitions between the primary activities are more fluid. The significant additions to the classic value chain are found particularly in the areas of skill management, life cycle management, IT support, knowledge generation as well as network setup and administration.

It is also interesting to note that sales shifts to the start of the value chain in solution business models. There is a shift away from *"We produce and then we sell"* to *"We configure a solution together with customers and implement it."*

As can be seen on the left of the figure, the changed value chain can also be subdivided into primary and support activities. This results in additional approaches for solution business models. In the area of support activities, these encompass the following fields (see Figure 36):

- Skill management
- Life cycle management
- IT support
- Network setup and administration
- Knowledge generation

Fig. 36 Example of a value chain changed through solution business models

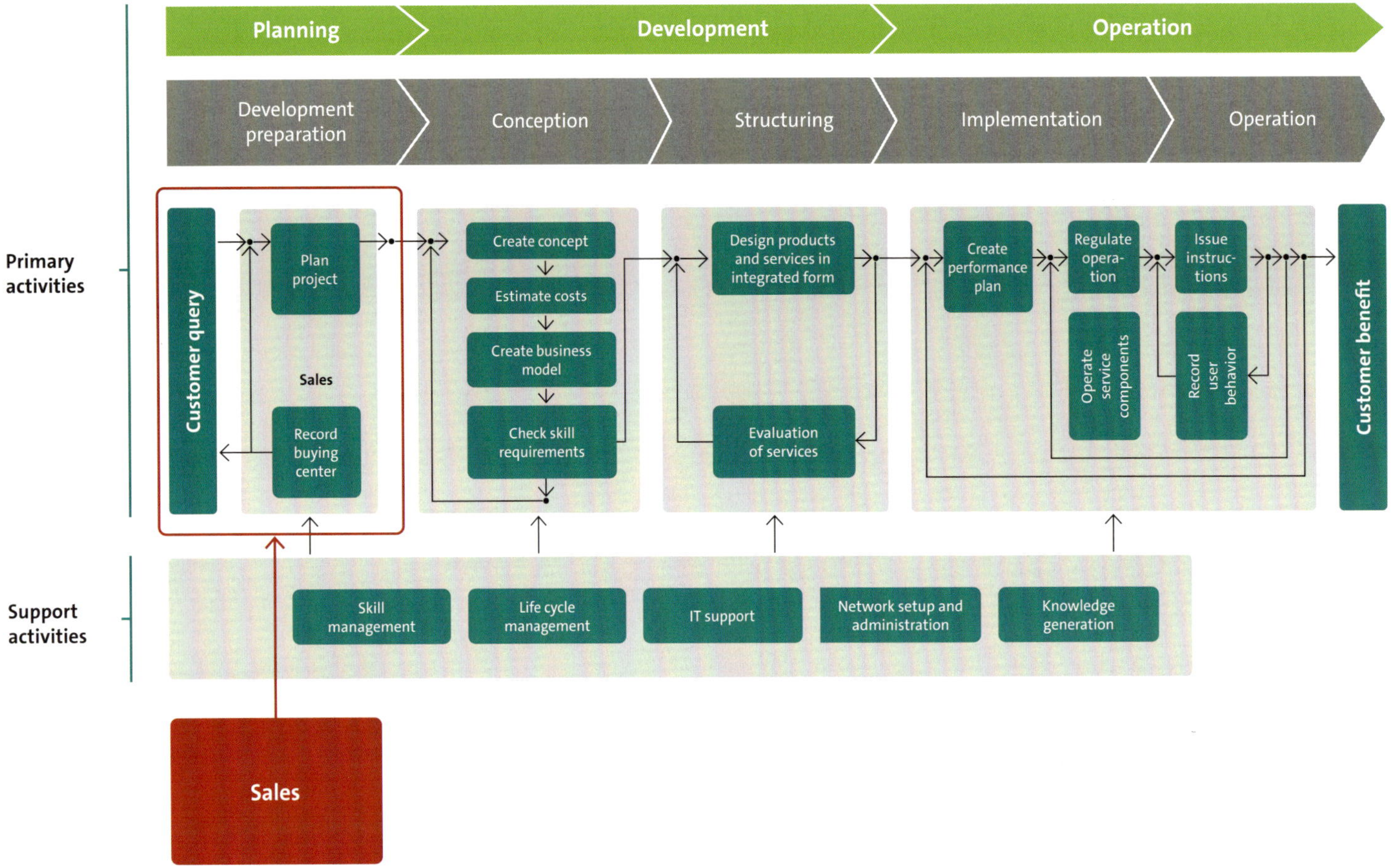

Figure based on: DFG collaborative research center, Transregio 29, TU Berlin, Ruhr University of Bochum: *Engineering Hybrid Service Bundles – Dynamic Interactions of Benefits in Kind and Services in Production.*

O – Operate processes for customers

Suppliers that penetrate deeper into the solution business models increasingly take over parts of their customers' value-creation processes. As shown in Figure 33 (p. 105), this increase from the 1st-step to the 2nd-step solutions is dramatic.

The logic lies in process efficiency: 2nd-step solutions and super-complex solutions are based on the assumption that the supplier is more efficient than the customer in the respective process.

Often, minor processes that take place upstream or downstream of the use of the supplier's products at the customer's site are initially taken over. The aforementioned *customer service* activities in *after-sales service* are the simplest form of taking over primary activities from customers. Just think of the comprehensive maintenance contracts that many companies conclude with their customers.

One example of taking over a supporting process in the value chain is provided by *Deutsche Lichtmiete (German Light Leasing)* from Oldenburg. In order to improve the *supporting infrastructure* of customer production, *Deutsche Lichtmiete* targets the lighting in production shops and storage facilities. The quality of the industrial LED lighting manufactured by the company (premium product) is supplemented by assuming responsibility for the lighting quality at the customer's site. *Deutsche Lichtmiete's* concept is indicated by the name of the company: Instead of selling lighting systems, they are leased to customers. In the process, lighting costs are permanently reduced and a lighting guarantee is provided throughout the entire lease period. The owner formulates the benefit of the solution as follows:

"We regard ourselves as a service company that takes care of selecting the appropriate products for each application area, light planning, installation, and lamp replacement for its customers – all for a pre-calculated rental price" (Marco Hahn, *Deutsche Lichtmiete*, impact magazine, March 2017, p. 19).

In advanced solution business models, entire process steps can be taken over by the supplier. Besides its steel products, for instance, steel producer *Salzgitter* has progressed to become a major railroad logistics company, and it undertakes railroad logistics for many of its customers. To do this, *Salzgitter* has built up its own network of regularly scheduled trains, called the *Salzgitter Logistics Train (SLoT). Salzgitter* describes this as follows: *"In contrast to usual individual wagon traffic with times of several days, today the SLoT system achieves transport times in the range of 24 hours. As a result, the coil is transported on a wagon with lower* CO_2 *emissions than on a truck."*[10]

In 2nd-step solutions and super-complex solutions, **operator models** often play an important role. The automotive industry is an example of this. As far back as the 1990s, *VW* had some processes of vehicle production in *VW* plants operated by its suppliers. Later on, the supplier *Magna* took over the entire production of the first-generation *BMW X3*. Today, operator models also play a role in many other industries such as the compressor and air technology, drive and materials handling technology, fluid technology, assembly, robotics, food and packaging technology, and machine tool sectors.

[10] https://www.salzgitter-flachstahl.de/de/news/artikel-der-salzgitter-flachstahl-gmbh/2016-02-24/lean-und-green-logistik-bei-der-salzgitter-flachstahl.html; accessed on 01.13.2021.

STRATEGY

O – Outcome-based pricing

Pricing also plays a role in distinguishing between the business model steps. In particular, *outcome-based pricing* is important. This refers to pricing models in which the price is no longer oriented to the manufacturing costs (cost-plus pricing), but to the value that arises for the customer thanks to the solution.

Put simply, it involves economy rather than manufacturing costs. Economy is oriented to the long-term relationship between the benefit achieved and the outlay (costs) required for it. This relationship is improved if the customer can continuously make its processes more effective and/or efficient thanks to the solution. One example is economy in the production of plastic containers for households. Here, automation solutions can increase both the quantity produced by the machines per hour and the efficiency in using the expensive polyethylene raw material.

Rolls-Royce's familiar "power by the hour" approach is usually referred to as the mother of all *outcome-based pricing* approaches. Instead of selling turbines to aircraft manufacturers, *Rolls-Royce* switched to "power by the hour" contracts. In this, the customer pays for each effective hour of turbine power but no longer owns the turbines itself.

In taking the step from product selling to 1st-step solutions, many suppliers start with leasing models. In these, the availability and permanent operational capability of devices, machines, and software solutions are guaranteed. 2nd-step and super-complex solutions primarily involve pricing models oriented to the result of the service. The manufacturing costs of the hardware (devices and machines) and software then play a subordinate role as regards the price. One example of this is the *Heidelberg Subscription* business model of *Heidelberger Druckmaschinen AG*. In this model, customers no longer pay for the machines themselves, but for their monthly printing volume.

However, it would be wrong to portray only the positive side of outcome-based pricing models because these models also pose one of the biggest risks for suppliers and can pave the way to Death Valley. That would be the case if the supplier's guarantees are too extensive or the performance calculations are too optimistic.

Outcome-based pricing models therefore necessitate a deep understanding of customer processes. To obtain this, the supplier requires empirical values concerning the longer-term course of the customer's product life cycles. If it does not have such empirical values, the supplier concludes contracts that are risky.

Let us look at the example of "power by the hour" again and put ourselves in *Rolls-Royce's* position. The pressure increases for *Rolls-Royce* to have the turbines up in the air because only then does it generate revenues. Therefore, the turbine technology has to be optimized time and again. But does *Rolls-Royce* not also have to understand its customers' business model better than if it simply sells the turbines? From the three business model questions in Chapter 3 (p. 77), let us just look at the question *"To whom is it marketed?"* Does it not make a difference if *Rolls-Royce's* airline customer wants to increasingly attract customers in the Middle East in the future? Then, the airline would be flying over desert regions in which sand could damage the turbines. If *Rolls-Royce* were unaware of this, it could lead to fatal miscalculations. Does *Rolls-Royce* not even have to predict the likelihood of the respective airline's pilots, cabin crew, and ground staff going on strike? That, too, would keep the turbines on the ground. And should even the statistical probability of pandemics such as COVID-19 be factored in?[11]

Hilti also has to have an in-depth understanding of its customers' processes to calculate the price of its fleet management. Due to its extensive guarantees, *Hilti* also has to be aware of the average costs of handling equipment unprofessionally. The possibility of the customers' employees intentionally damaging machines also has to be factored in; after all, *Hilti* guarantees to exchange defective equipment, irrespective of the reason. If this knowledge were not available, *Hilti* would calculate the pricing model too optimistically.

Consequently, the requirements for sales change: It is no longer sufficient for companies that operate with *outcome-based pricing* to be roughly familiar with their customers. Instead, companies such as *Rolls-Royce* and *Hilti* have to penetrate to the heart of their customers' operating conditions, processes, applications, and problems.

STRATEGY

[11] More information about the risks of *Rolls-Royce's* business model can be found in: Plötner (2016), *Cost traps in business models 4.0*, ESMT Berlin, p. 8.

F – Far-reaching relationship

Another significant characteristic that distinguishes the classic product business and solution business models is the longevity and intensity of the relationship between the supplier and the customer. To date, the relationship with the customer in many product-centric business models is transactional in nature. This means that there is little interaction between the supplier and customer between purchasing and potential repurchasing. In evolving to become a solution provider, relationships with customers become more long-term and intensive.

As shown in Figure 33 (p. 105), this SNOOFY factor usually increases sharply between the 1st- and the 2nd-step solutions. However, even simple solutions such as after-sales services result in more points of contact between suppliers and customers over a longer period of time than in the pure product business.

Far-reaching also means that considerably more people are involved on both the supplier side and the customer side. On the customer side, it is no longer only the purchasing department and the users that are involved in the purchase decision, but also additional functions (such as IT experts) and hierarchy levels (such as the management). This affects the structure of the sales team (see Chapter 13, p. 313).

Often, *far-reaching* also means that the supplier's networking has to extend beyond the customer company. Just think of the aforementioned "Building Information Modeling (BIM)" solutions in major construction projects. An entire "BIM elephant round" is needed here, meaning a network consisting of the client, developer, scaffolding and concrete formwork suppliers, installation companies, electrical contractors, pipelayers, and utility companies, among others.

An important distinction in the form of the relationship between the supplier and customer can be determined between operator models and Industry 4.0 business models during the transition from 2nd-step solutions to super-complex solutions. Looking at operator models, you could be forgiven for thinking that the trend toward outsourcing is continuing apace. In actual fact, the essential change in progressing from 2nd-step solutions to super-complex solutions no longer lies in "operate *for* customer," but in "operate *with* customer." Figure 37 traces this development. Taking over customer processes no longer increases, whereas integration with the customer does. This is often referred to as "co-creation" by suppliers and customers. In this process, the suppliers and their employees are integrated deeply into the customer company over a specific period of time. This ranges from joint supplier/customer teams up to and including integrated production plants.

Fig. 37 The difference between "operate *for* customer" and "operate *with* customer"

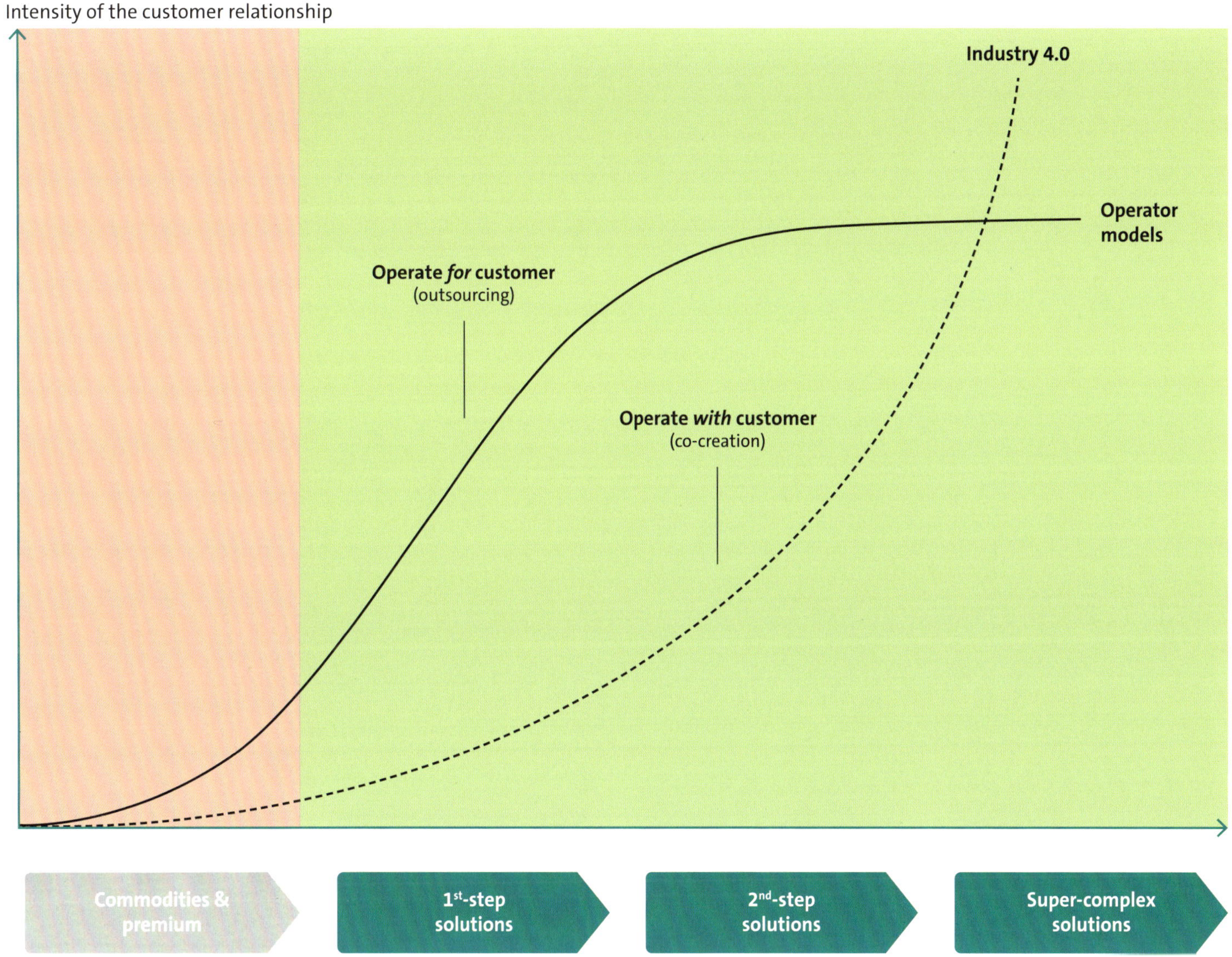

Y – Yield customer value-in-use

The SNOOFY factor ***Yield customer value-in-use*** now involves the questions of **when** and **how** the benefit is created.

Let us first take a look at the question of **when**. In the logic of the classic product business, the value of the product can be easily accounted for at the point in time of delivery. Conversely, value-in-use is concerned with the value that occurs throughout the entire application or usage duration of a product or service. It therefore looks at the "benefit during usage." Instead of supplying the customer with a product with an optimal price/quality ratio at the time of the transaction, a longer-term perspective is now involved. The aim is to increase the value of the hardware and software during the course of usage.

This temporal orientation is shown in the center of Figure 38. While the purchase time point lies after the product has been manufactured in the case of commodity and premium products, significant parts of service provision only take place afterwards in the solution business model steps. This future orientation involves a **risk** for the customer. It is based on the difficulty of estimating the future benefit, the possibility of being dependent on the supplier, and, in the worst case, having to pay more than originally thought in the long term.

In order to counteract this customer risk, solution providers offer *guarantees*. Such guarantees can extend to various areas. Often, temporal availabilities or specific machine output quantities per unit of time are guaranteed.

The figure illustrates this development using the example of *PERI*. *PERI's* development began with the pure supply of formwork and scaffolding for buildings, but it has expanded to include the acceptance of more extensive responsibility, with the result that *PERI* is now a *value-in-use* company.

Fig. 38 Value-in-use through increasing acceptance of responsibility by the supplier

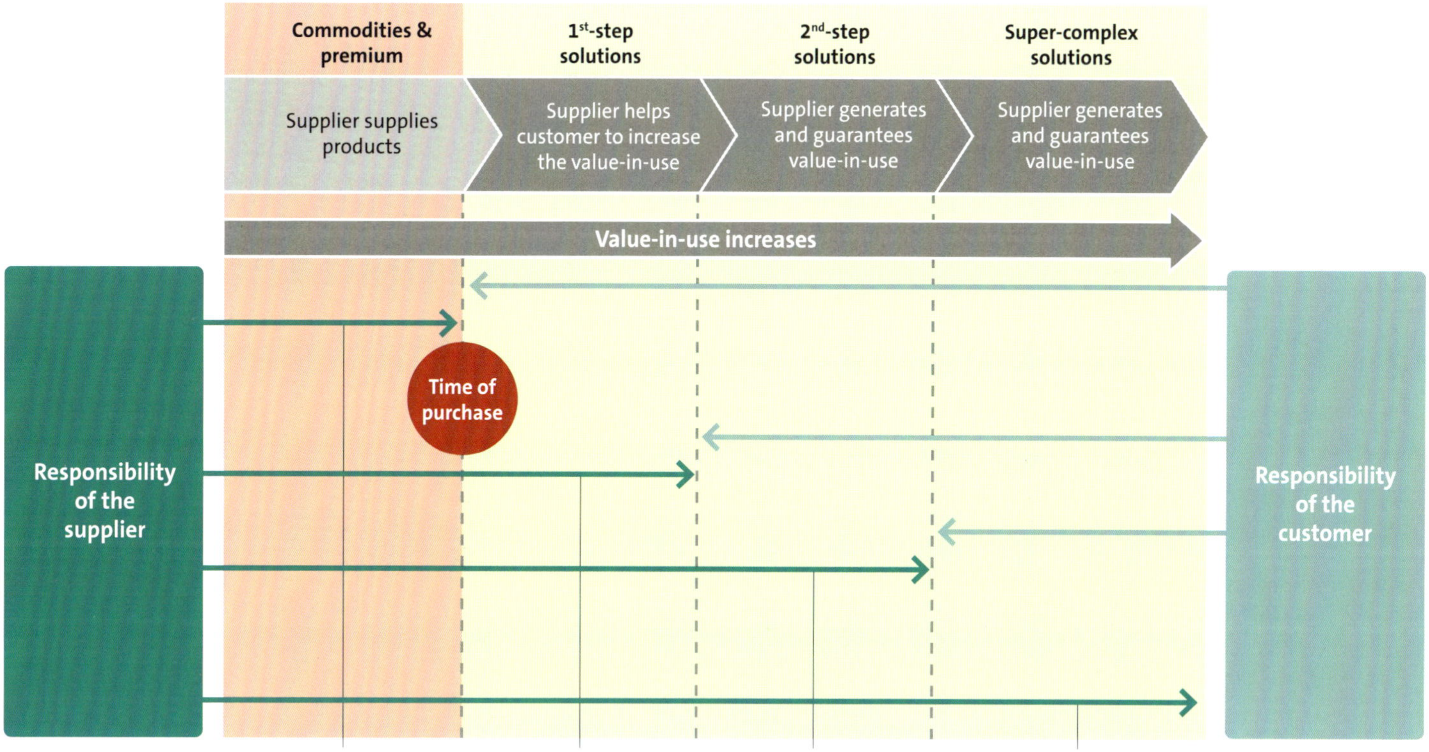

The supplier sells a product to the customer.

Example
PERI supplies formwork and scaffolding to its customers.

The supplier helps the customer to optimize its processes.

Example
PERI advises its customers so they receive the optimal material for each specific building.

The supplier assumes responsibility for processes in the customer's value chain.

Example
PERI takes over project management processes to guarantee the optimal use of personnel and material on construction sites.

The supplier assumes responsibility for results.

Example
PERI guarantees value-in-use, e.g., it ensures that the project costs do not exceed a specific level.

Let us now address the question of ***how** the benefit is created*. The answer lies in the responsibility assumed by the supplier. In Figure 38, this is symbolized by the range of the arrows between the supplier and customer. In shifting toward solution business models, the value-in-use rises due to the increasing acceptance of responsibility.

- In product-centric business models (commodity and premium), the supplier assumes no responsibility for the value-in-use. The focus here is on supplying products with which the customer continues to work independently after purchasing. Conversely, solution providers search for possibilities to increase the value for customers during the usage process.
- In 1st-step solutions, suppliers provide their customers with initial assistance in increasing their value-in-use. On this business model step, the **benefit during usage** can be increased, for example, through consulting services or simple process improvements, as in the case of *Michelin* (see information box).
- On transition to 2nd-step and super-complex solutions, the supplier assumes even greater responsibility. Here, the supplier not only accepts responsibility for usage readiness (in the sense of a machine that is ready for use), but also for the result of usage. This means that guarantees are provided for production quantities and times, or for prices and process costs.

Offers aimed at the customers' value-in-use usually reflect this assumption of responsibility in the pricing system. In this regard, there is a logical link between ***O**utcome-based pricing* and ***Y**ield customer value-in-use*. There is also a relationship between ***N**urture/**O**perate customer processes* and ***Y**ield customer value-in-use* because supporting and taking over customer processes are ultimately aimed at increasing the customer value-in-use.

Focusing on value-in-use instead of the optimization of product facets often necessitates a longer phase of rethinking. The effects on structures (see Chapter 13), leadership (see Chapter 11), and skills (see Chapter 9) in sales are radical.

Information box: *Michelin Solutions* as an example of increasing value-in-use

Michelin Solutions is part of the *Michelin Group*. With around 800 employees, this business unit looks after more than 300,000 vehicles in vehicle fleets throughout Europe. *Michelin Solutions* pursues the objective of improving the efficiency and economy of vehicle fleets of all kinds.

Three solution concepts are offered to accomplish this:[12]

1. Optimized trailer use with the aid of digital control systems
2. Outsourcing solutions for tire management with guaranteed fuel economy
3. Digitally networked tire services for fleets that want to manage their tires themselves using efficient tools

Michelin Solutions enters into a contractual commitment to achieve a reduction in fuel consumption or the number of flat tires that is defined together with the customer.

[12] Source: https://www.michelin.de/news-center/Michelin-solutions-praesentiert-massgeschneiderte-Fuhrparkloesungen-auf-der-transport-logistic; accessed on 03.19.2018.

The SNOOFY business model test

In our scientific work, we have validated a test method that can be used to determine a SNOOFY score. Both a total score and also a separate score for each of the six SNOOFY factors are determined in this process. The business model test consists of a total of 32 questions (see following pages).

You can perform the SNOOFY business model test yourself. To do this, feel free to contact us:
sales-management-department@ruhr-uni-bochum.de.

In our evaluation system, we first use industry values to validate individual business model criteria. We then use a conversion factor to calculate the values of the individual SNOOFY factors and the total SNOOFY score.

The value ranges of the total SNOOFY score for the three solution business model steps are as follows:

- 100–124: 1st-step solution
- 125–139: 2nd-step solution
- 140–150: Super-complex solution

In Chapter 7, I present 18 application examples in the form of business model profiles. In these, I portray both the total SNOOFY score and the profile of the individual SNOOFY factors according to the principle of Figure 39.

Fig. 39 The online questionnaire for the SNOOFY business model test

The business model test helps to answer the following questions:

- On which business model step are you located with your current and planned new solution offers?
- Industry benchmark: Where do your individual business models lie compared to the average in your industry?
- Which of the SNOOFY factors offer the opportunity for differentiation from the competition?

The total SNOOFY score enables the classification of business models into the business model steps. In turn, the business model step then impacts the strategy and sales:

SALESTEGY

- **Strategy**: Decisions regarding the strategic positioning of one's own business models (see Chapter 6)
- **Sales:** Decisions regarding the entire spectrum of sales management (see Chapters 9–14)

Three notes on correctly classifying the SNOOFY score:

1. **The test cannot replace a more in-depth analysis.** I do not want to raise any false hopes at this point. Our business model test can prove very helpful for an initial classification. However, it is not a substitute for dealing with the business model in greater depth.
2. **The level of the SNOOFY score does not equate to a positive or negative evaluation of the business model.** Instead, the aim is to be able to assess the business model step on which a business model is located. As mentioned above, classification into a business model step is not per se positive or negative (see p. 94).
3. **It remains a continuum.** Classification into one of five business model steps must not obscure the fact that all of the SNOOFY criteria involve a continuum. The transitions are fluid.

Information box: Scientific validation of the SNOOFY business model test with 845 companies

We validated the SNOOFY factors individually and as a whole using reliability and factor analyses. All statistical quality indicators comply with the current scientific reliability and validity criteria. The internal consistency of the individual measurement scales therefore lies above the critical Cronbach alpha value. The exploratory and confirmatory factor analyses also meet the additional quality criteria for determining the convergence and discriminant validity.

The following table shows the evaluation indicators used in the SNOOFY business model test. As can be seen, the test consists of a mix of subjective assessments and objectifiable variables.

The SNOOFY business model test for classifying business models

Instructions: The SNOOFY business model test should be performed separately for each of your business unit's business models. Therefore, focus on one of your business models when answering the following questions.

Servitization										
In this business model, we essentially act as a ...										
... producer of products/hardware.	−2	−1	0	+1	+2	... service provider.				
							Disagree completely			Agree completely
This business model is focused on services.							1	2	3	4	5
In this business model, we generate our main margins with services, not with products/hardware.							1	2	3	4	5
In this business model, the customer benefit arises through the service, not through the product/hardware.							1	2	3	4	5
In this business model, everything is oriented to services, not our products/hardware.							1	2	3	4	5
How high is the service content of the business model's sales in %?*							Benchmark with industry Ø				

Nurture customer processes: Supporting customer processes											
Cobbler, stick to your trade. In this business model, we are focused very traditionally on our *product quality* and not on the customer processes.	−2	−1	0	+1	+2	In this business model, we do not differentiate ourselves through our product quality, but through the *improvement of customer processes*.					
							Disagree completely				Agree completely
In this business model, we see ourselves as process facilitators for our customers.							1	2	3	4	5
The service offer supports our customers in improving their processes.							1	2	3	4	5
We advise our customers on how they can improve their processes.							1	2	3	4	5
We are not only focused on optimizing our own production. We are primarily concerned with optimizing customer processes.							1	2	3	4	5

Operate processes for customers: Taking over customer processes						
In this business model ...						
... we supply products and/or advise our customers, but we do not take over part of the customer processes ourselves.	−2	−1	0	+1	+2	... we actively intervene in the customer value chain and take over parts of the customer processes.

	Disagree completely				Agree completely
In this business model, we take over *supporting processes* in the customer value chain (e.g., tool management or maintenance).	1	2	3	4	5
In this business model, we take over *core processes* in the customer value chain (e.g., production or logistics in their entirety).	1	2	3	4	5
In this business model, we offer our customers an operator model in which we operate, e.g., machines, systems, or production plants for the customers.	1	2	3	4	5
In this business model, we offer to take over a complete process for our customers, i.e., they outsource an entire value-creation process to us (e.g., complete production, complete logistics, complete marketing).	1	2	3	4	5

Outcome-based pricing:						
In this business model, the price to be paid is oriented ...						
... to our manufacturing costs (cost-plus pricing).	−2	−1	0	+1	+2	... to the value that arises for the customer (outcome-based pricing).

	Disagree completely				Agree completely
The price of this service offer arises from the output that is generated (e.g., per manufactured quantity, per hour of service).	1	2	3	4	5
In this business model, we have continuous income (e.g., monthly invoices) that goes beyond a one-off purchase.	1	2	3	4	5
The price to be paid is dependent on the benefit (economy) to the customer.	1	2	3	4	5
The price is oriented to the result that the customers achieve (e.g., number of products manufactured in a day by a machine).	1	2	3	4	5

Far-reaching relationship						
In this business model, there is …						
… no intensive interaction with the customers beyond the purchase process.	−2	−1	0	+1	+2	… cross-departmental, close cooperation with the customers over a long period of time.

	Disagree completely				Agree completely
In this business model, we undertake "co-creation," i.e., we work intensively together with the customers across various departments and functions.	1	2	3	4	5
The interaction intensity between us and the customer company is very high, both before and after the purchase.	1	2	3	4	5
The specific solution only arises through close cooperation with the customer. Customer feedback is continually used to adapt the solution.	1	2	3	4	5
In this business model, there are permanent project teams whose members come both from the customer and from us.	1	2	3	4	5
What is the average duration of a customer relationship in this business model?*	Benchmark with industry Ø				

Yield customer value-in-use						
In this business model, we are focused on ...						
... the value of our offer at the time of purchase. This therefore involves the optimal price/performance ratio.	−2	−1	0	+1	+2	... the value that arises for the customer throughout the entire application/usage duration. It therefore involves the "benefit during usage."

	Disagree completely				**Agree completely**
In this business model, we assume extensive responsibility (e.g., in the form of guarantees) for the long-term benefit of our solution offer to the customer.	1	2	3	4	5
In this business model, we do not think in terms of transaction to transaction, but rather of the quantifiable benefit that is intended to arise for our customers over a matter of years.	1	2	3	4	5
In this business model, the satisfaction of our customers is oriented not only to the quality of a supplied product, but also to the benefit that we enable throughout the entire usage duration.	1	2	3	4	5
How well the value-creation processes function at the customer's site is our responsibility. To meet this responsibility, we attempt to extensively understand the customers' value-creation processes in this business model.	1	2	3	4	5

* Comparison with the industry values is necessary for this question.

WHY SALESTEGY?

STRATEGY – FINDING THE RIGHT BUSINESS MODELS

PRACTICAL EXAMPLES

SALES – IMPLEMENTING NEW BUSINESS MODELS IN SALES

Chapter 6
The SNOOFY business model audit

- Step 1: Defining the target direction of the SNOOFY business model audit
- Step 2: Performing the SNOOFY business model test for classifying business models
- Step 3: Analyzing business model dynamics in your market
- Step 4: Identifying and avoiding Death Valleys in business models
- Step 5: Re-evaluating competitive advantages

Message of this chapter

The SNOOFY business model audit can be used to conduct a strategic analysis of the business models.

What should you have learned by the end of the chapter?

- The SNOOFY business model audit consists of five steps.
  - Step 1 is concerned with defining the target direction of the audit: Should it be used to analyze new or existing business models?
  - In step 2, the SNOOFY business model test is applied to determine the steps on which your business models are located and to evaluate how you are positioned compared to the industry benchmark.
  - In step 3, the dynamics of your market are analyzed to prevent you from being caught out by disruptive developments.
  - Step 4 is concerned with anticipating Death Valley risks that can arise when establishing new business models.
  - In step 5, customer and competitor segments are analyzed, and future market opportunities are derived on this basis.

DEATH VALLEY MARKETS

Strategy of minimum costs and prices

GREEN HILL MARKETS

Strategy of solution business models

FINDING

Finding solution business models

CHAPTER 3

Reducing complexity

CHAPTER 4

Distinguishing between five business model steps

CHAPTER 5

Using SNOOFY factors for differentiation

CHAPTER 6

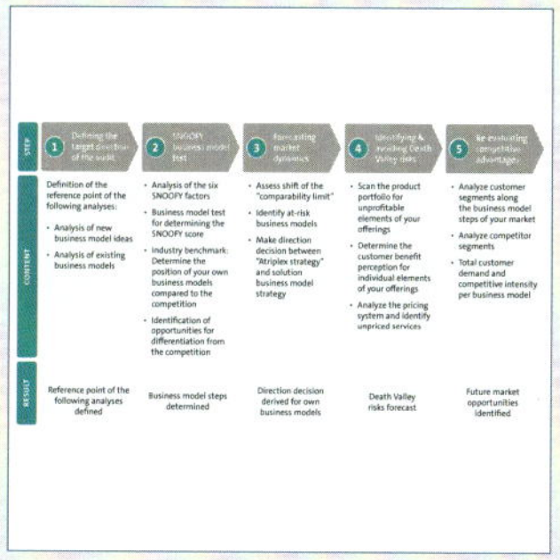

Audit: Identifying opportunities for competitive advantages in the five steps

STRATEGY

The SNOOFY business model audit consists of five steps. The procedure is shown in the following figure.

Fig. 40 The five steps of the SNOOFY business model audit

STEP	1 Defining the target direction of the audit	2 SNOOFY business model test	3 Forecasting market dynamics	4 Identifying & avoiding Death Valley risks	5 Re-evaluating competitive advantages
CONTENT	Definition of the reference point of the following analyses: • Analysis of new business model ideas • Analysis of existing business models	• Analysis of the six SNOOFY factors • Business model test for determining the SNOOFY score • Industry benchmark: Determine the position of your own business models compared to the competition • Identification of opportunities for differentiation from the competition	• Assess shift of the "comparability limit" • Identify at-risk business models • Make direction decision between "Atriplex strategy" and solution business model strategy	• Scan the product portfolio for unprofitable elements of your offerings • Determine the customer benefit perception for individual elements of your offerings • Analyze the pricing system and identify unpriced services	• Analyze customer segments along the business model steps of your market • Analyze competitor segments • Total customer demand and competitive intensity per business model
RESULT	Reference point of the following analyses defined	Business model steps determined	Direction decision derived for own business models	Death Valley risks forecast	Future market opportunities identified

Step 1: Defining the target direction of the SNOOFY business model audit

The first step is concerned with defining the target direction of the SNOOFY business model audit. There are two fundamental target directions (see Figure 41):

- Analysis of new business model ideas
- Analysis of existing business models

The first case usually involves a detailed analysis of individual business models. The second case is particularly important for companies operating with various business models on the market. For companies pursuing five or more business models, this involves the portfolio strategy.

Fig. 41 Defining the target direction of the audit

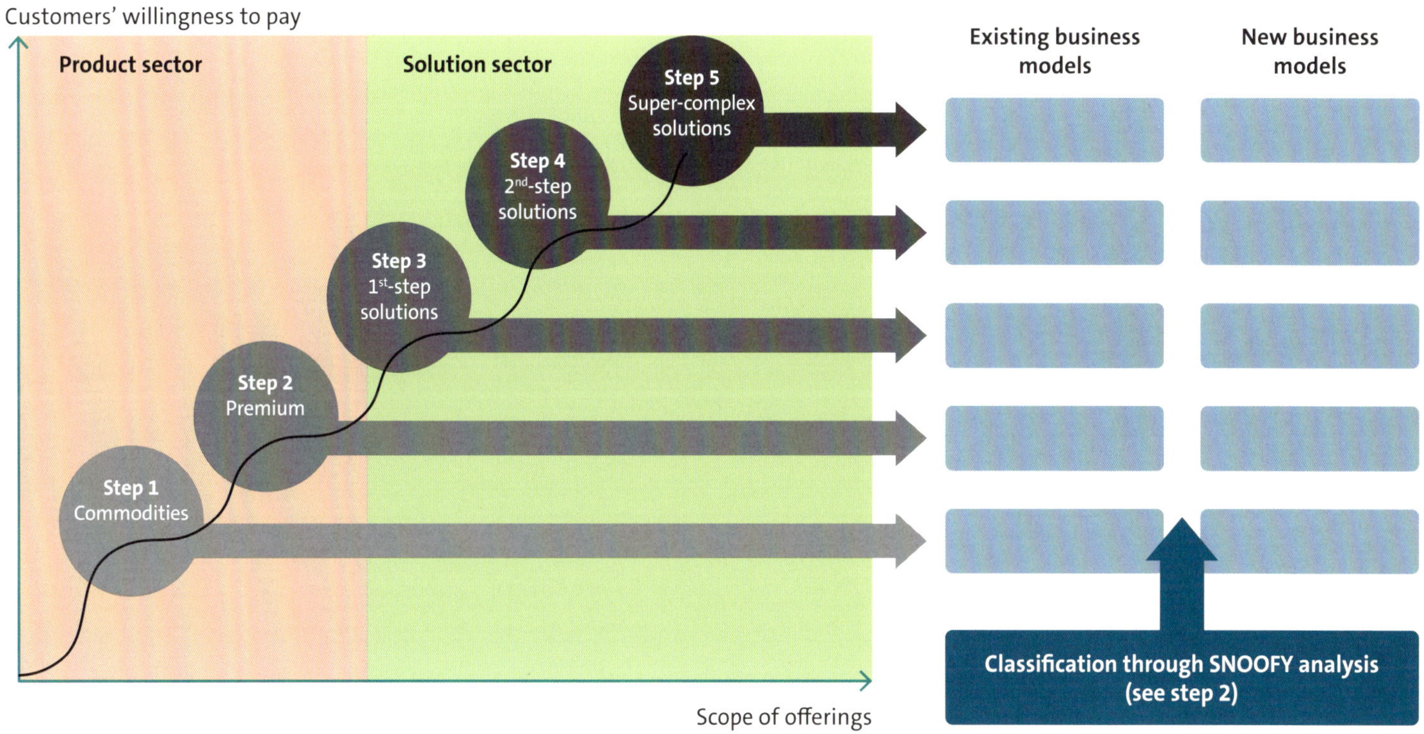

Step 2: Performing the SNOOFY business model test for classifying business models

The second step involves classifying the business models (see Figure 42). The SNOOFY business model test can be used to help you do this.

Fig. 42 The five steps of the SNOOFY business model audit

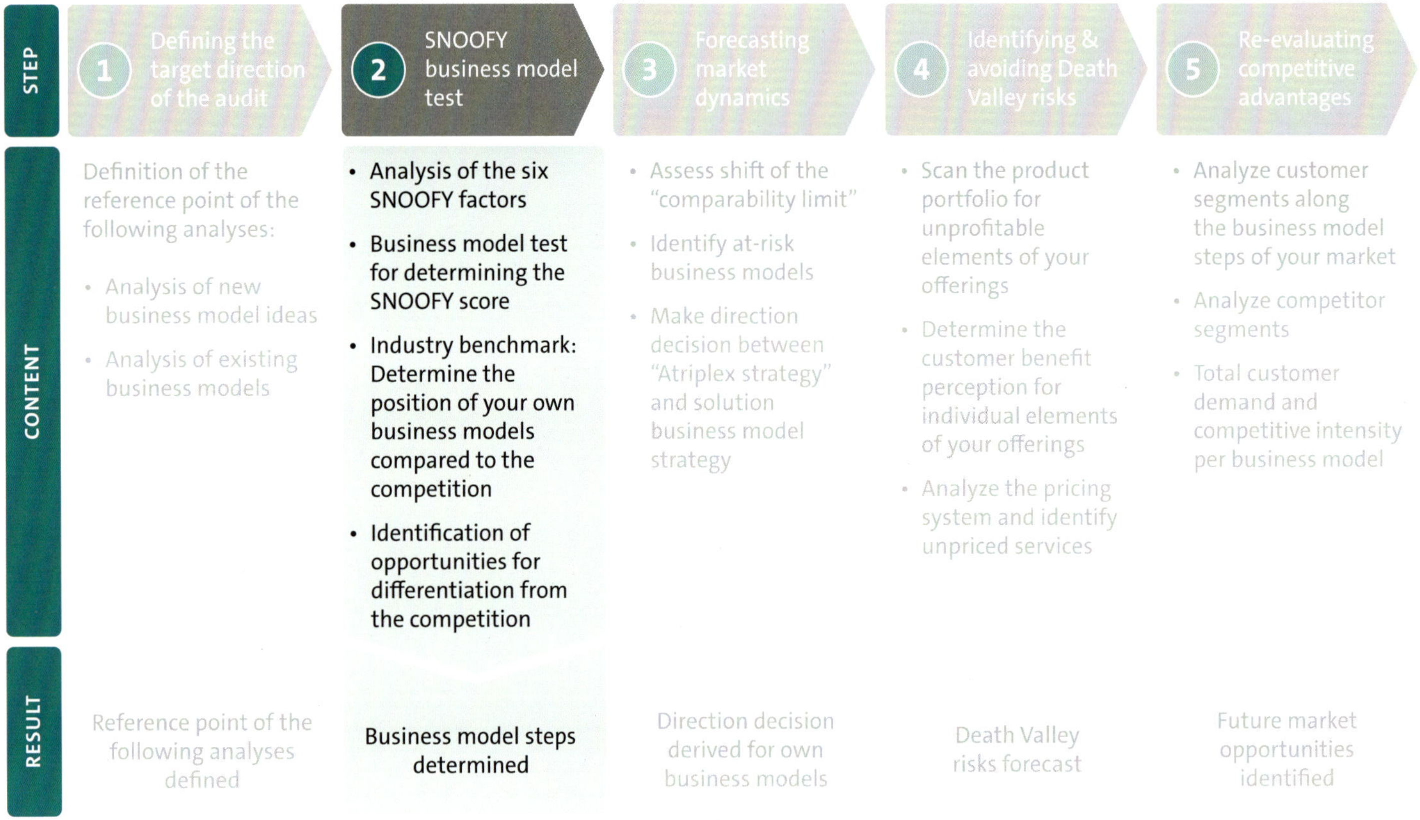

The six SNOOFY differentiation factors can be used to determine the step to which a business model belongs. The following questions have to be answered in the overall assessment using the SNOOFY factors:

- On which business model step are we located with our current and newly planned solution offers?
- Industry benchmark: Where do our individual business models lie compared to the average in our industry?
- Which of the SNOOFY factors offer the opportunity for differentiation from the competition?

The answers to these questions are important because they enable you to draw the right conclusions concerning the optimal structural organization and control of your sales department. You also require the comparison with the industry benchmark to be able to review your own competitive position. In doing so, you have to take account of the special considerations of your industrial branch. Due to its nature, for example, the service content (**S**ervitization) on all business model steps is significantly higher in the IT industry than in other industries. To do this, we have set up a business model database at our department so that you can benchmark your business models.

The *SNOOFY business model test* described in Chapter 5 for determining the SNOOFY score is used as an initial aid in this process. The test enables you to determine both the total SNOOFY score and the values of the individual SNOOFY factors.

Step 3: Analyzing business model dynamics in your market

An understanding of the market dynamics is required to be able to assess the sustainability of business models (step 3 of the SNOOFY business model audit, see Figure 43). This is all the more true in the digital age. What is called the **"comparability limit,"** which I now explain in detail, plays a crucial role in this.

Fig. 43 The five steps of the SNOOFY business model audit

STEP	1 Defining the target direction of the audit	2 SNOOFY business model test	3 Forecasting market dynamics	4 Identifying & avoiding Death Valley risks	5 Re-evaluating competitive advantages
CONTENT	Definition of the reference point of the following analyses: • Analysis of new business model ideas • Analysis of existing business models	• Analysis of the six SNOOFY factors • Business model test for determining the SNOOFY score • Industry benchmark: Determine the position of your own business models compared to the competition • Identification of opportunities for differentiation from the competition	• Assess shift of the "comparability limit" • Identify at-risk business models • Make direction decision between "Atriplex strategy" and solution business model strategy	• Scan the product portfolio for unprofitable elements of your offerings • Determine the customer benefit perception for individual elements of your offerings • Analyze the pricing system and identify unpriced services	• Analyze customer segments along the business model steps of your market • Analyze competitor segments • Total customer demand and competitive intensity per business model
RESULT	Reference point of the following analyses defined	Business model steps determined	Direction decision derived for own business models	Death Valley risks forecast	Future market opportunities identified

The comparability limit identifies the level of complexity up to which the offers in a market can be easily compared by means of platforms, online retailers, and comparison portals. Beyond this limit, customers are only able to compare the offers precisely with a great deal of effort.

The comparability limit therefore determines whether a business model already lies within the range of high market transparency or whether this is to be anticipated in the near future. A participant at one of our management seminars once paraphrased this as follows: *"I see that as a weather front between Death Valleys and green hills."*

Let us use a simplified market portrayal to visualize the market dynamics (see Figure 44). Let us assume that it shows the transitions from the commodity to the premium product business and to the 1st-step solution business.

As you can see on the left, the comparability limit at time point 1 lies at the start of the commodity business. At time point 2, it lies at the start of the premium product business. At time point 3, the comparability limit has shifted so far to the right that almost all premium product offers are already in the transparency range. Beyond the comparability limit, however, the offers are so complex that hardly any comparison is still possible. The proportion of customizable service elements that are adapted to customer processes is so high that the complexity is too great for online platforms.

Fig. 44 The dynamics of the comparability limit

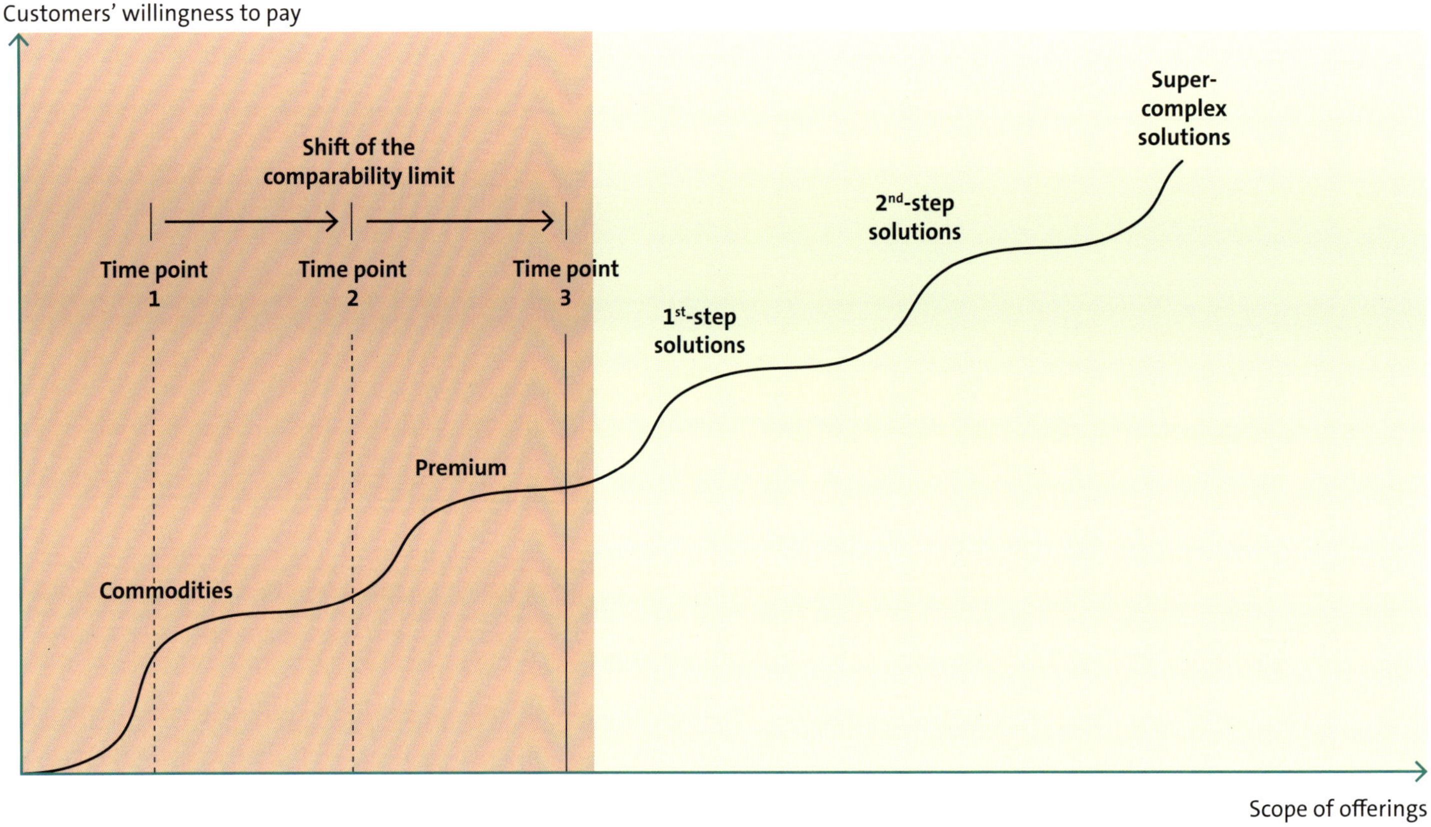

The increasing options of digitalization play a central role in the development of the comparability limit. Whereas we were previously only able to compare very simple products at the far left, the transparency of prices and performance parameters is increasingly shifting to the right due to digitalization. Many complex premium goods can now be found on comparison portals. This also applies to services with multiple parameters, such as insurance, trade services, and moving services.

Figure 45 shows the next phase of market dynamics after the comparability limit has shifted. The price level in the *commoditized area* drops considerably. This reflects the fact that many customers rationally decide in favor of the least-expensive supplier in transparent market sectors with identical offer parameters.

Fig. 45 Price development after the comparability limit has shifted

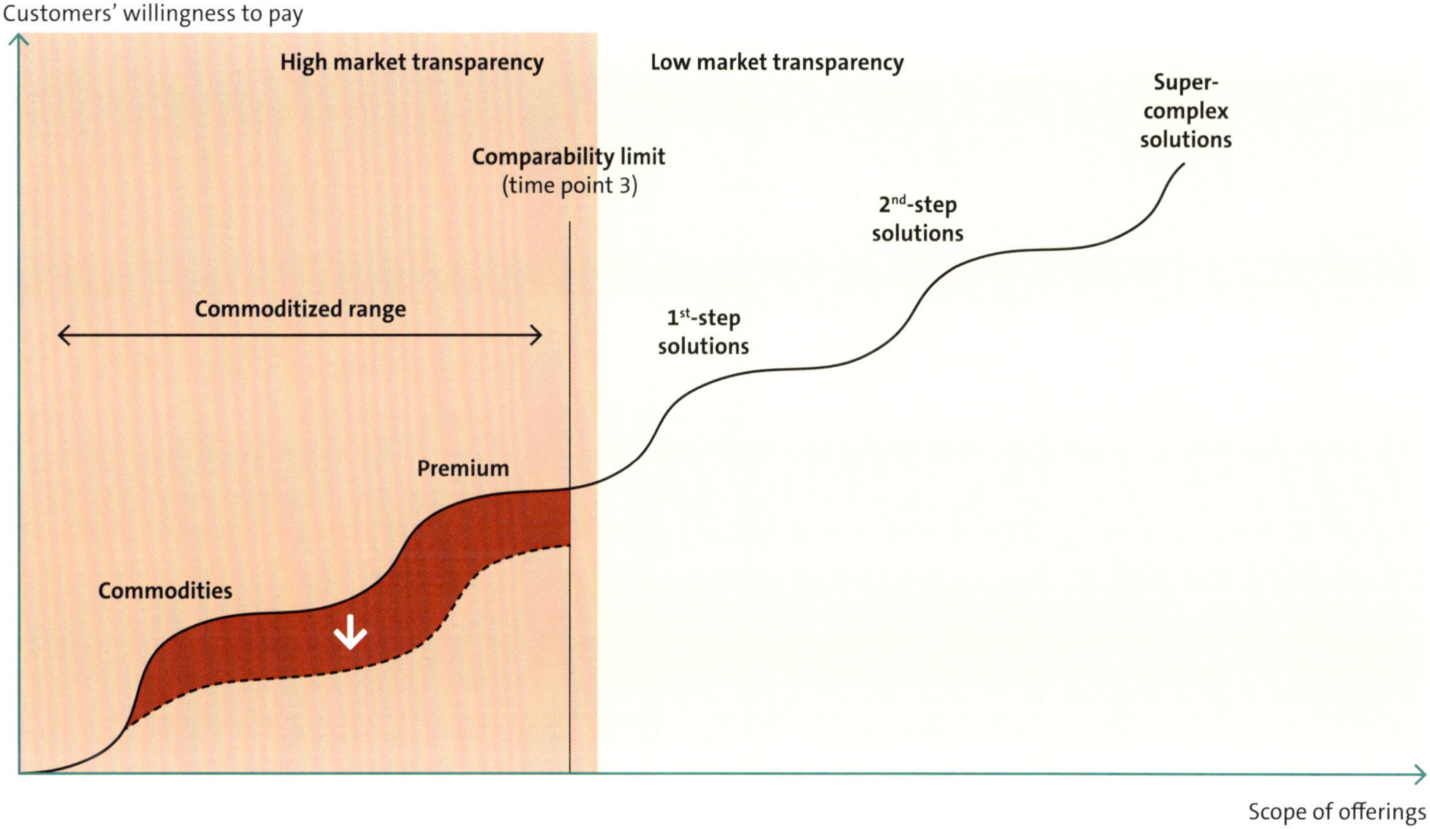

When and how quickly the comparability limit shifts depends on the capabilities of the digital platforms. Over time, offers that were long considered to not be digitally comparable are overtaken by the comparability limit. This can be referred to as increasing *commoditization* because this transparency leads to a breakdown of the barriers to switching on the part of customers.

Due to these market dynamics, the market analysis must take the following question into consideration: *What can be commoditized next in your market due to digital platforms?*

It is worthwhile performing a simple analysis to determine just how close the comparability limit has already come for individual business models. Figure 46 shows a helpful tool for this. It visualizes which offer areas were still difficult for customers to compare three years previously but are now transparent today. It also forecasts which offers might be overtaken by the comparability limit in three years.

Fig. 46 Forecasting commoditization in the market

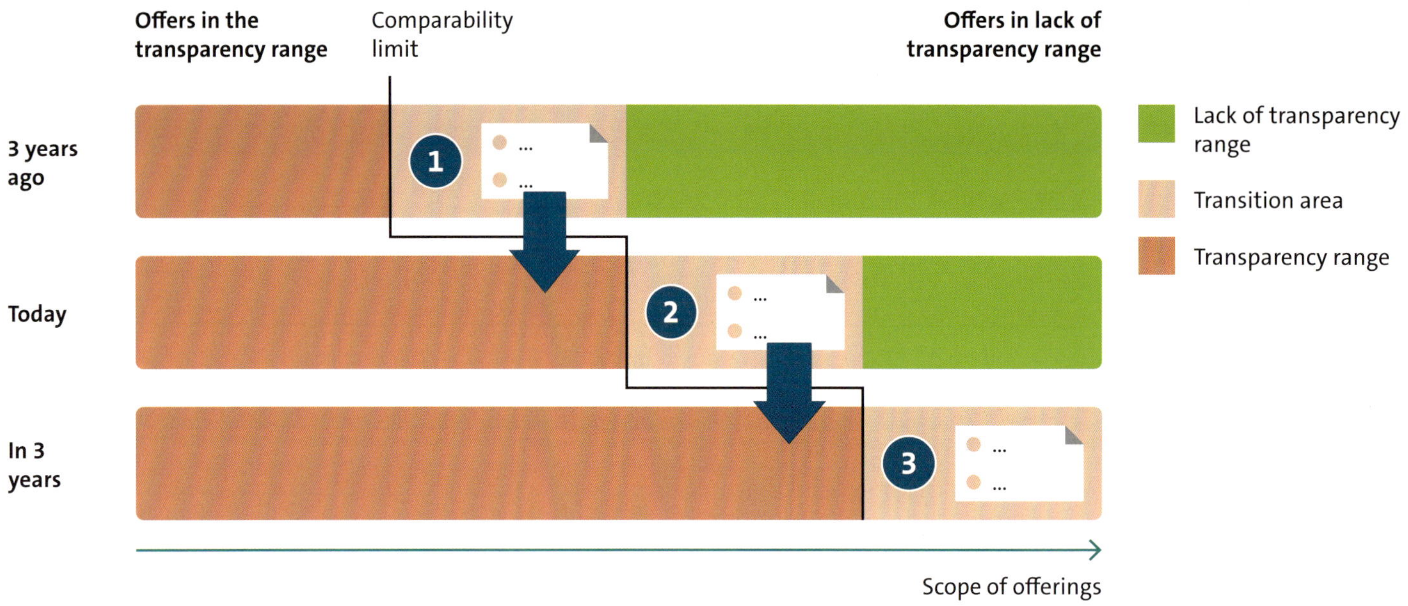

Which products/services in our market ...

1. ... were still in the lack of transparency range three years ago but are now comparable using digital platforms?
2. ... are still in the lack of transparency range today but will presumably be comparable in three years?
3. ... will presumably still be in the lack of transparency range in three years?

The following figure shows an example from the banking market for business customers. The results were developed by the managers of a major bank. In the upper line, it can be seen that many areas of the previous, classic core business, such as current accounts, investment loans, and some parts of foreign exchange trading, now lie in the transparency range.

The most fascinating area for the major bank managers was the second line concerning the question of what is likely to shift into the transparency range in the next few years. In their opinion, this range included employer-financed company pensions and public funding programs.

In order to correctly forecast the changes in market transparency in these product areas, it is necessary to assess both the next drivers of digitalization (particularly in the area of artificial intelligence and the development of cooperations with *FinTech platforms*) and state banking regulations (such as the mandatory disclosure of individual offer elements).

Fig. 47 Commoditization of a market using the example of business customer banking

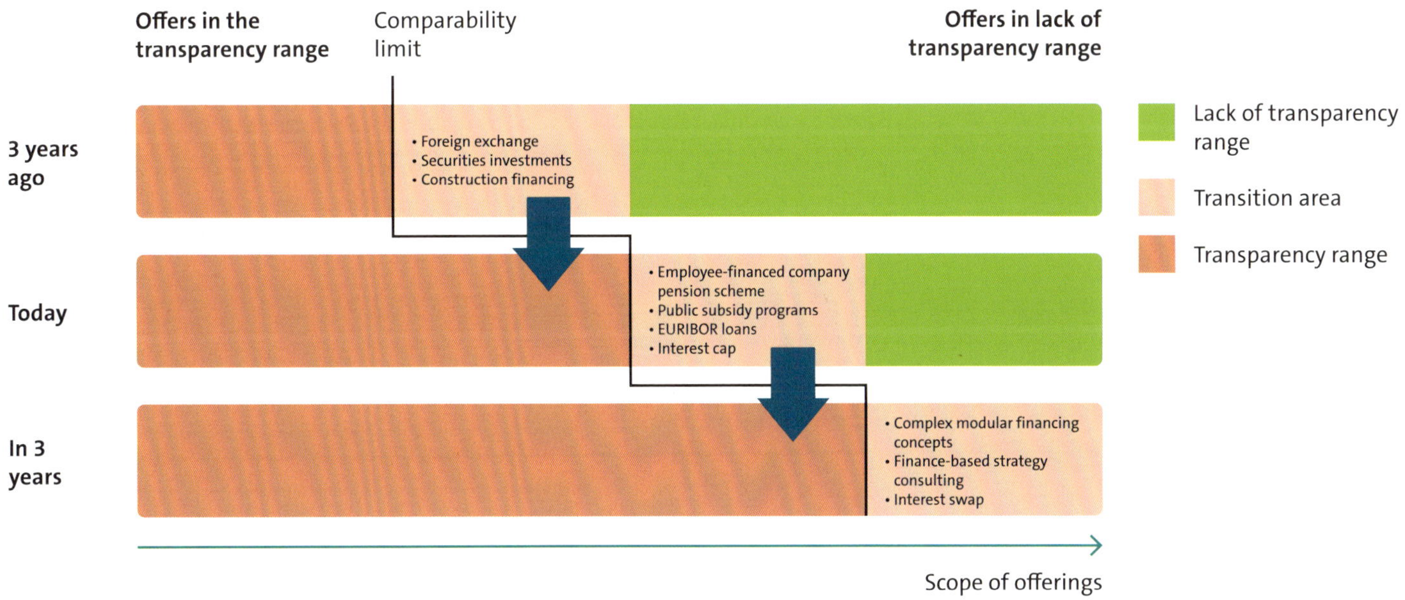

"In which direction should we move?" In light of market dynamics, we are often asked this question by management decision-makers. There are two fundamental movement directions (see Figure 48).

The first leads to the commodities. An *Atriplex strategy* is consistently pursued in this case. This is the strategy of maximum efficiency similar to that of the desert plant that we encountered in Death Valley (see Chapter 1, p. 41).

Drastic measures that not only encompass production costs, but also minimize the costs of all other value-creation processes, such as the costs for research and development, logistics, marketing, and sales, are needed to do this on the cost side. Only this enables constant price leadership in commodities.

Fig. 48 Two movement directions for business models in dynamic markets

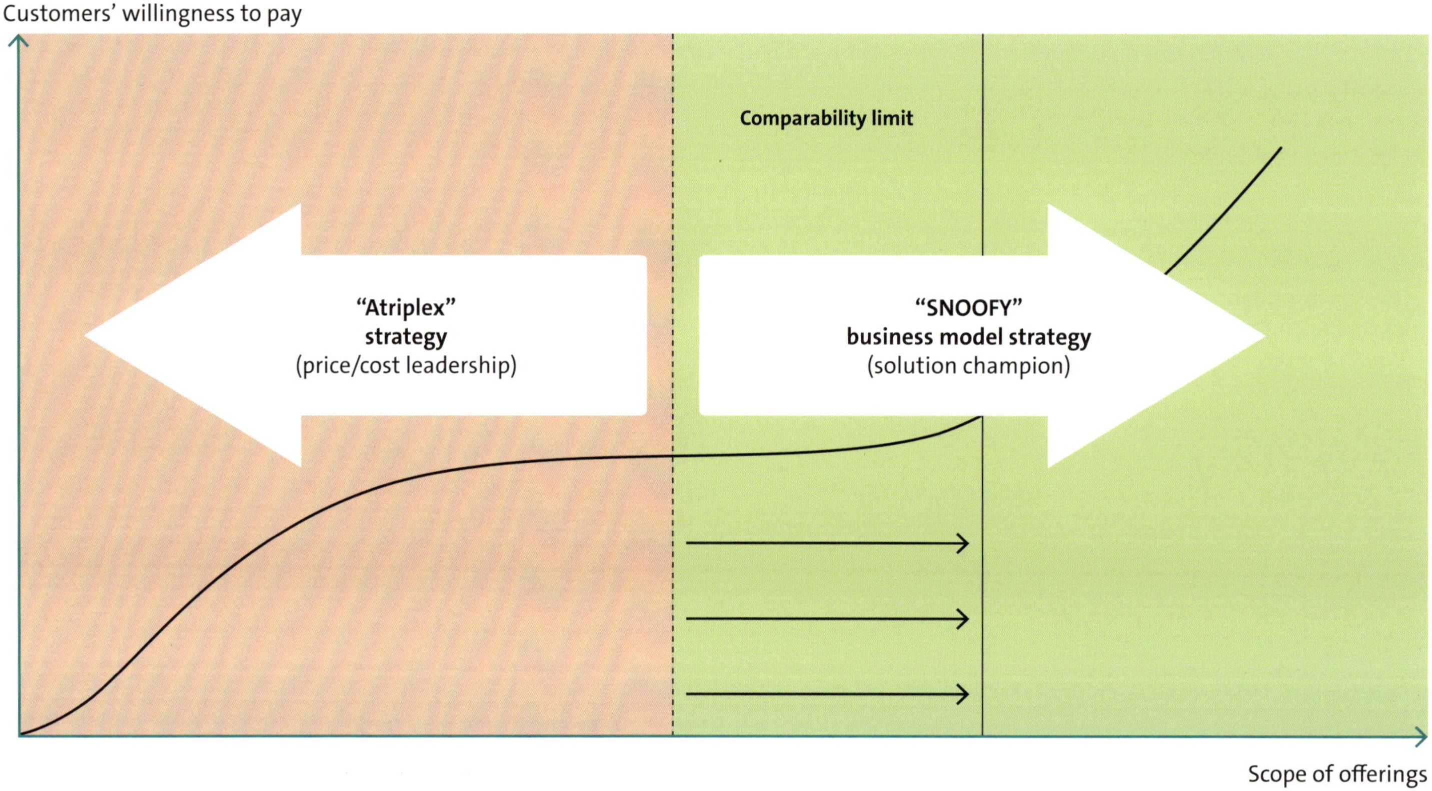

Atriplex strategies are minimal-cost business model approaches. Offer elements that are not central to the target group are forgone. Examples in the B2C sector include *Motel 6*, and low-cost airlines such as *Southwest Airlines*. *Siemens* and *Dow* are pursuing this strategy in the B2B sector (see information box).

Therefore, the Atriplex strategy is also being pursued by suppliers from highly developed industrialized nations. Olaf Plötner deals exhaustively with this in *Counter Strategies in Global Markets* (2012, Palgrave Macmillan) under the concept of "no-frills technologies."

Despite these examples, the Atriplex strategy does *not* embody the usual competitive advantage of B2B suppliers from highly developed industrialized nations. The wage costs and regulatory requirements there are too high for that. Consequently, the total revenue share of "SMART" products at *Siemens* is also minimal.

Information box: Two examples of Atriplex business models

"Xiameter" – the Atriplex business model of *Dow*

The business model of chemicals group *Dow's* silicone division was based on 1st-step solution offers. In this, the sale of silicone products was paired with extensive advice. However, *Dow* gradually slid into a Death Valley, as some of its customers were increasingly less willing to pay for this advice. These customers were merely interested in the sale of the product at the best price. *Dow* responded to this with an Atriplex strategy. In parallel with the previous business model, a commodity offer aimed solely at low costs was created under the *Xiameter* brand. *Xiameter* is sold via the online channel, which is limited to standardized products.

"SMART" – the Atriplex business model of *Siemens*

At *Siemens*, the Atriplex strategy was implemented under the acronym "SMART." *Siemens* describes this minimalistic business model as follows:[13]

"SMART stands for simple, maintenance-friendly, affordable, reliable, time to market. 'Siemens has doubled its SMART product portfolio in China in the last two years and we definitely have to do more,' says Rückl [Head of Detector Center – Computed Tomography at Siemens Healthineers]. Knowledge gained from the Chinese market is now also flowing into solutions for other markets, such as India and Brazil. CT scanners from China are also sold in these countries today."

STRATEGY

[13] https://www.siemens.com/innovation/en/home/pictures-of-the-future/research-and-management/innovation-management-ct-china.html; accessed on 03.19.2018.

The second action option involves penetrating the complexity of the SNOOFY business model factors. The road to 1st-step, 2nd-step, and super-complex solutions is aimed at positioning oneself beyond the comparability limit with the objective of avoiding transparency and the brutal price competition in product markets.

If we take another look at the five business model steps of highly developed markets, we can recognize **digitalization in its second role** (see Chapter 3, p. 76), in which it enables entirely new solutions. Suppliers can use these to penetrate into market sectors that are not very transparent due to their complexity.

The risk of excessively large leaps in business model development. Our observations show that a disproportionately high number of companies fail if they attempt to skip business model steps. For instance, the probability of failing when jumping from the product business to 2nd-step solutions or even super-complex solutions is far higher than if a company has already acquired experience in the 1st-step solution business.

Many companies pay dearly when making the bigger leap, often into uncharted territory. One example of this is the super-complex solution *"Soarian"* from *Siemens* (see p. 152). If a company nevertheless decides to skip business model steps, it has to factor in a large time and budget safety margin.

Let us sum up. The following core questions should be answered when analyzing the business model dynamics:

1. How will the comparability limit develop in your market? What could be *commoditized* next in your market due to digitalization?
2. Can you establish the prerequisites for an *Atriplex strategy* in order to become a cost leader that can survive even in transparent Death Valley markets? This means: Can you establish structures for minimum wage and process costs in research and development, purchasing, production, sales, and logistics?
3. If not, which options do you have for developing solution offers?
4. Does this pose a risk of taking excessively large business model leaps? If you would like to skip business model steps, have you factored in a sufficient time and budget safety margin for unforeseen implementation difficulties?

Step 4: Identifying and avoiding Death Valleys in business models

The following statements indicate that companies are in a Death Valley with their solution offers. The fourth step of the SNOOFY business model audit involves being able to identify such Death Valleys in business models and thereby avoid them (see Figure 49).

- *"Many customers' willingness to pay for our additional services is low. They regard them as a free add-on for ensuring customer loyalty."* (Sales manager of a logistics company)

Fig. 49 The five steps of the SNOOFY business model audit

STEP	1	2	3	4 Identifying & avoiding Death Valley risks	5
CONTENT	Definition of the reference point of the following analyses: • Analysis of new business model ideas • Analysis of existing business models	• Analysis of the six SNOOFY factors • Business model test for determining the SNOOFY score • Industry benchmark: Determine the position of your own business models compared to the competition • Identification of opportunities for differentiation from the competition	• Assess shift of the "comparability limit" • Identify at-risk business models • Make direction decision between "Atriplex strategy" and solution business model strategy	• Scan the product portfolio for unprofitable elements of your offerings • Determine the customer benefit perception for individual elements of your offerings • Analyze the pricing system and identify unpriced services	• Analyze customer segments along the business model steps of your market • Analyze competitor segments • Total customer demand and competitive intensity per business model
RESULT	Reference point of the following analyses defined	Business model steps determined	Direction decision derived for own business models	Death Valley risks forecast	Future market opportunities identified

- *"We're constantly improving our service offers to our customers, but their willingness to pay doesn't increase anything like in proportion to the costs that we incur for it."* (Sales director of a B2B wholesaler)
- *"Our machine is now around 30 units per hour faster and also offers around 21% higher resolution (...) and we're still treading water in terms of sales volume, gross margin, and market share."* (Sales director of a mechanical engineering company)

Fig. 50 Each business model step also involves a Death Valley risk

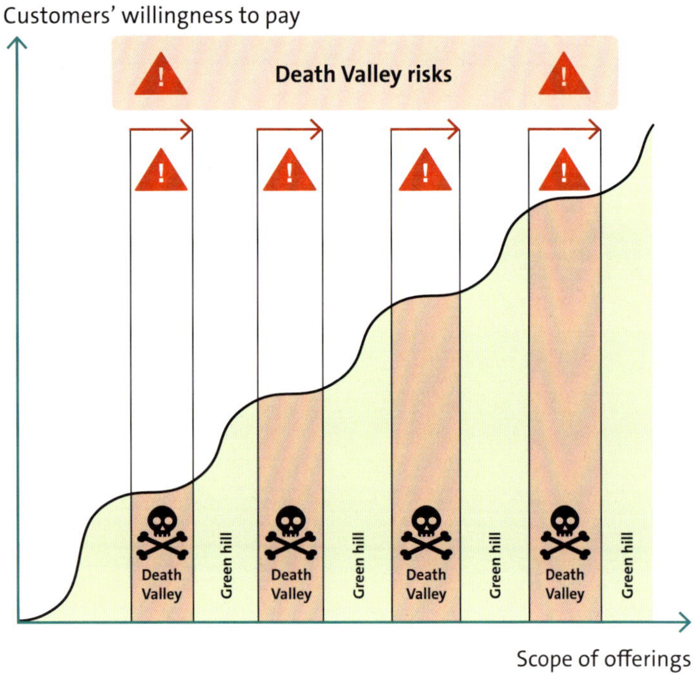

The Death Valleys are already shown in the business model step figure. Let us take another look at the basic figure of the five business model steps. One central finding of our work is that this figure conceals several Death Valleys that we have not yet shown. Only the following figure makes them visible. They are hidden in the flat transition areas between the business model steps (Figure 50).

Let us look again at the nature of Death Valley. There is something there that we have disregarded so far: death zones at various altitude levels. Death Valley's lowest level is situated in *Badwater*, a depression that lies 85.5 meters below sea level. On the way out of Death Valley, however, there are other barren intermediate plateaus that also reveal desert-like habitat conditions. For instance, *Saline Valley* (300 meters above sea level) and *Racetrack Playa* (1,120 meters above sea level) are situated higher up in Death Valley than Badwater. They also exhibit extremely hostile desert conditions. Having escaped one hazardous Death Valley level does not mean that there are no others to be encountered at a higher level.

Accordingly, new business model steps not only offer companies the positive characteristics of green hills – a sharp increase in the willingness to pay. **Each new business model step also poses a Death Valley risk!** As Figure 50 shows, Death Valleys can occur on **each** business model step. With every scope of offerings comes a risk of unprofitability zones.

In the Death Valley areas, the additional services are not compensated by a sufficient willingness to pay on the part of customers. As can be seen in the figure, the scope of offerings increases there but customers' willingness to pay does not. These effects correspond to the basic definition of Death Valleys in Chapter 1 and demonstrate that suppliers are only able to sell their products and services with very low profit margins in these market sectors. After all, additional services cost the suppliers money. However, these expenses are not balanced with increased income.

Definition of Death Valleys in the business model portfolio:
Death Valleys in the business model portfolio are unprofitable performance aspects of an offering. The supplier increases the scope of its offerings, but the costs of this are not compensated by a sufficient willingness to pay on the part of the customers. These aspects of an offering have to be identified in a business model audit.

Death Valleys are entirely natural during the transition to the next business model step! At first glance, you could be forgiven for assuming that additional solution steps lead to substantial profits for the suppliers. However, a closer look shows that almost all companies have had to contest with Death Valleys on their way to the next step. This also applies to the business model examples in Chapter 7. Even shining examples initially ended up in Death Valleys. *Rolls-Royce*, too, occasionally generated high losses with its power-by-the-hour business model.

3x new territory. Death Valleys are entirely natural on the road to new business model steps because new ground is being trod in three areas (see following table):

1. ***New customer understanding territory.*** New customer benefit areas are addressed with the new solution offers. That necessitates a new dimension of customer understanding. This encompasses an increasingly deeper understanding of the customer value chain.

   This customer understanding should also include the risks perceived by the customers because if a customer is intended to place parts of its value chain in the hands of others, this necessarily leads to the relinquishment of power and control. This results in the risk of hardly being able to intervene if the supplier fails to provide a solution or behaves opportunistically. This fear of dependency and the possible costs of switching makes many customers skeptical. A lot of customers shy away, particularly from data-based solutions, because they are afraid that the supplier might access sensitive information.

   The crucial skill in sales is therefore being precisely familiar with customer requirements and risks. To this end, we conducted research under the term "customer need knowledge" (CNK). The management must have this CNK for each target customer segment and the sales employees must have it for each individual customer.

2. ***New pricing territory.*** New pricing becomes necessary due to the new solution offers. Even if no fundamental pricing system switch (from cost-plus pricing to outcome-based pricing, for example) is taking place yet, the supplier must price the additional service elements of the offer adequately.

   The first challenge in this case lies in sales. Customer habit stands in the way of the willingness to pay for a new scope of offerings.

Many customers are still accustomed to the fact that new services are not priced separately. The sales department therefore has to determine the right sales methods to convey the monetary benefit of the offer and the related price to the customers. I discuss the sales methods in greater detail in Chapter 14.

The second challenge lies on the strategic level. It involves adequately covering all of the costs and risks on the supplier's side in the pricing system. This means, for example, that the fixed costs for reserving staff for maintenance contracts have to be factored into the price just like those for provisions due to extensive guarantees.

3. ***New sales territory.*** The new service offers necessitate a new sales setup (see Chapters 8–14).

   On the other side of the coin, there are often long-established sales structures, management and compensation approaches, and familiar sales methods. However, these are often doomed to failure when moving into new business model steps. For example, sales structures involving a high visit frequency, hard selling approaches, and high variable compensation may function well in the product sector, but such an approach would be unsuitable in the solution business.

   Each business model step requires a sales setup tailored precisely to the specific requirements.

Death Valley check-up

New territory	Death Valley risks	Check-up questions	Recommendation
1. New customer understanding territory Experience regarding which benefits and which risks customers perceive in the new business model steps is not yet available.	From the customers' point of view, the new service package does not offer sufficient benefit. • Sales fails to determine the customers' individual, complex needs. • The customers' value chains have not yet been penetrated sufficiently. • The salesforce is not yet familiar with the most important benefit aspects and the biggest purchase risks per customer segment.	• Have you analyzed the most important benefit aspects in the customer value chains for your target customer segments? • If you select three customers at random, can the responsible account manager tell you precisely what the most important benefit aspects are for these customers? • Is the account manager familiar with the risks due to which these customers would advise other customers against purchasing?	• Adapt sales methods to obtain a deeper understanding of customers (more in Chapter 14). • Carry out a benefit/risk check on a representative selection of customers.

Death Valley check-up			
New territory	**Death Valley risks**	**Check-up questions**	**Recommendation**
2. New pricing territory Experience of pricing the new business model steps is not yet available.	The set price is not selected correctly for the new solution package. • It does not adequately represent the benefit to the customers, and/or • it does not adequately represent the supplier's costs.	• Are you familiar with the characteristics in the solution package that engender the greatest willingness of customers to pay? • Do you regularly determine whether individual characteristics are becoming more or less relevant for customers? • Have you conducted a scenario analysis (best, middle, worst cases) for your long-term cost development for the new business model step?	• Establish a pricing radar for each business model step. • Calculate and track cost scenarios. • Conduct regular surveys to determine customers' willingness to pay for individual aspects of the offering.
3. New sales territory Experience of the sales structure for and the implementation of new business model steps is not yet available.	• The skills, structures, and leadership approaches in sales do not yet match the new solution package. • Sales is not able to communicate the offer successfully.	• Have you checked whether your current sales setup meets the requirements of the planned business model step? • Have you defined a process for the cooperation between research and development, sales, and marketing?	• Gear up sales for the requirements of the respective business model steps in all areas. The project steps in Chapters 9–14 can be used for this. • Establish a permanent interface between research and development, sales, and marketing.

Cerner Soarian can be used as an example of Death Valley risks when moving into new business models (see information box). *Cerner* bought *Soarian* from *Siemens* for $1.5 billion in 2014.

This case conceals a Death Valley disaster that took place during the initial years before this business model was adopted because the road to this highly complex hospital information system was embarked on **without a sufficient focus on sales**. The *Soarian* case stands as an example of many other cases in which a lack of focus was placed on sales when establishing new business models.

Information box: Death Valley risks of *Cerner Soarian*

Cerner offers comprehensive software packages for hospitals under the brand name *Soarian*. They can be classed as a *super-complex solution* because they map highly complex core hospital processes. *Cerner* refers to *Soarian* as "workflow engines" that support hospital process management. *Soarian* uses various modules to help hospitals optimize and centrally control processes in virtually all areas. This particularly includes patient management, in which each patient is represented in the *Soarian* workflows, thus enabling each process step to be tracked. This extends from administration of the patient file containing all diagnoses up to and including outpatient medication. Unnecessary waiting times, inefficiencies, and errors (e.g., medication errors) are minimized in this way.

At the same time, *Soarian* is also an example of how a company entered a Death Valley when introducing a new business model.[14] *Soarian* was originally introduced by *Siemens*. *Siemens* offers a long-term wealth of experience in the healthcare sector with an array of products. These products include various high-tech devices in the area of medical imaging, such as computed tomography (CT) scanners and magnetic resonance imaging (MRI) scanners. Based on its observations that market competition was increasing even in premium product segments, *Siemens* launched *Siemens Soarian* onto the market just after the turn of the millennium.

But what caused *Siemens Soarian* to fail, with the result that *Siemens* sold the division to *Cerner*? And why was *Cerner* able to implement the *Soarian* business model profitably in the market?

One of the main reasons was sales. *Siemens* started out with a sales budget that was far too low. The management also overlooked the fact that selling *Soarian* posed other requirements for the sales employees compared to selling classic CT and MRI products. Although CTs and MRIs are also expensive, purchasing *Soarian* represented an entirely new dimension for hospitals and, accordingly, the enormity of the decision was far greater. All of the hospitals' governing bodies had to be convinced of the solution, something that could not be accomplished with the existing salesforce. The requirements for sales were so demanding that the HR department was barely able to recruit the right employees. Looking back, Tom Miller, the former boss of *Siemens Soarian*, formulated this self-critically as follows: *"People from the traditional business did not even have the right skills to interview the necessary candidates."*[15]

When it bought *Soarian*, *Cerner* focused explicitly on the sales requirements of the business model.

[14] A detailed description of the *Soarian* case can be found in: Plötner, Habel, Schmitz (2020), *Complex Service Solutions*; in: *Solid Growth – Global Strategies for Industrial Champions*, BTM Center, ESMT Berlin, pp. 7 ff.

[15] Quoted from Plötner (2016), *Cost traps in business models 4.0*, ESMT Berlin, p. 14.

Death Valley risk due to overoptimistic expectations. *"The application's 18 new functions will ensure that we cover all customer segments and needs."* Such overoptimistic expectations are a Death Valley risk for companies moving toward solution business models. What if the costs of developing the new application are not recovered through orders? What if the biggest competitor has developed an application that only includes three new functions, but precisely these are of most importance to customers?

Besides overestimating the customer benefit and customers' willingness to pay, the following are also often overestimated or underestimated when entering solution business models:

- **Internal (time) requirements and costs are underestimated.** The requirements that a change of business models makes on sales and the amount of time that this change takes are often underestimated. The lean period until new business model steps start to generate money can be long. As a result, the necessary sales budget and also the change costs are often underestimated.
- **Income is overestimated.** According to our investigations, the fact that demand for – and thus the income from – new solution business models is overestimated poses a major problem. This is particularly dramatic as regards demand in the initial phase. This form of miscalculation proves fatal because complex solution business models, in particular, usually take some time to be amortized.

  One of my interview partners aptly remarked: *"Ninety percent of business cases for innovative business models overestimate income in the startup phase. There usually are no further phases because the negative cash flow leads to an abrupt end beforehand."*
- **Risks of new pricing models are underestimated.** Changing the pricing model is the special risk of new solution business models. I dealt with the switch to outcome-based pricing models in the description of the fourth SNO**O**FY factor ***(Outcome-based pricing)***. This switch is rational from the value-in-use-oriented perspective of the solution business. At the same time, many suppliers run risks here, since they are not fully aware of the ramifications of their new pricing model. In effect, outcome-based pricing models equate to assuming part of the risk previously borne by the customer.

Step 5: Re-evaluating competitive advantages

In the fifth and final step of the SNOOFY business model audit, the competitive advantages thanks to new business models are assessed and future market opportunities are derived (see Figure 51).

Fig. 51 The five steps of the SNOOFY business model audit

STEP	1 Defining the target direction of the audit	2 SNOOFY business model test	3 Forecasting market dynamics	4 Identifying & avoiding Death Valley risks	5 Re-evaluating competitive advantages
CONTENT	Definition of the reference point of the following analyses: • Analysis of new business model ideas • Analysis of existing business models	• Analysis of the six SNOOFY factors • Business model test for determining the SNOOFY score • Industry benchmark: Determine the position of your own business models compared to the competition • Identification of opportunities for differentiation from the competition	• Assess shift of the "comparability limit" • Identify at-risk business models • Make direction decision between "Atriplex strategy" and solution business model strategy	• Scan the product portfolio for unprofitable elements of your offerings • Determine the customer benefit perception for individual elements of your offerings • Analyze the pricing system and identify unpriced services	• Analyze customer segments along the business model steps of your market • Analyze competitor segments • Total customer demand and competitive intensity per business model
RESULT	Reference point of the following analyses defined	Business model steps determined	Direction decision derived for own business models	Death Valley risks forecast	Future market opportunities identified

If the comparability limit in a market shifts and the business models come under pressure as a result, one's own competitive advantages have to be reviewed. The managing director of a quality leader in mechanical engineering for metal production described this as follows: *"Previously, the competitive advantages were distributed relatively simply. Along with two to three competitors, we were the clear quality leaders, while a couple of others had established themselves as price leaders. Because of complex solution offers, it is much more difficult to define your own competitive advantage today."*

The final step in the SNOOFY business model audit therefore encompasses a check-up of your own competitive advantages. This enables you to determine which business model steps you should move toward in the future.

To do so, you can proceed in three analysis steps. The first two steps involve an analysis of the customer segments (A) and an analysis of the competitor segments (B). These two analyses are conducted both for the current market and for developments over the next few years. On this basis, you can then determine the future competitive advantage potentials (C).

A. Analysis of customer segments: Which customers demand which business model step?

In the B2B sector, customer segmentation is often carried out simply on the basis of customer size and target industries. Essentially, this is a helpful approach. Nevertheless, we also have to analyze customer segments with reference to the development of the business models. After all, there are entirely different customer segments for the five business model steps in highly developed markets.

Let us go through the most frequently found customer segments using the example of a market with five business model steps. Naturally, there are various patterns for subdividing the customer segments for such markets. A frequently seen segment pattern is shown in Figure 52. In the boxes above the business model steps, the respective demand is indicated by the brightness level. The darker the box, the greater the customer segment demand for the business model step. Each line reflects one of the customer segments.

Fig. 52 Analysis of customer segments

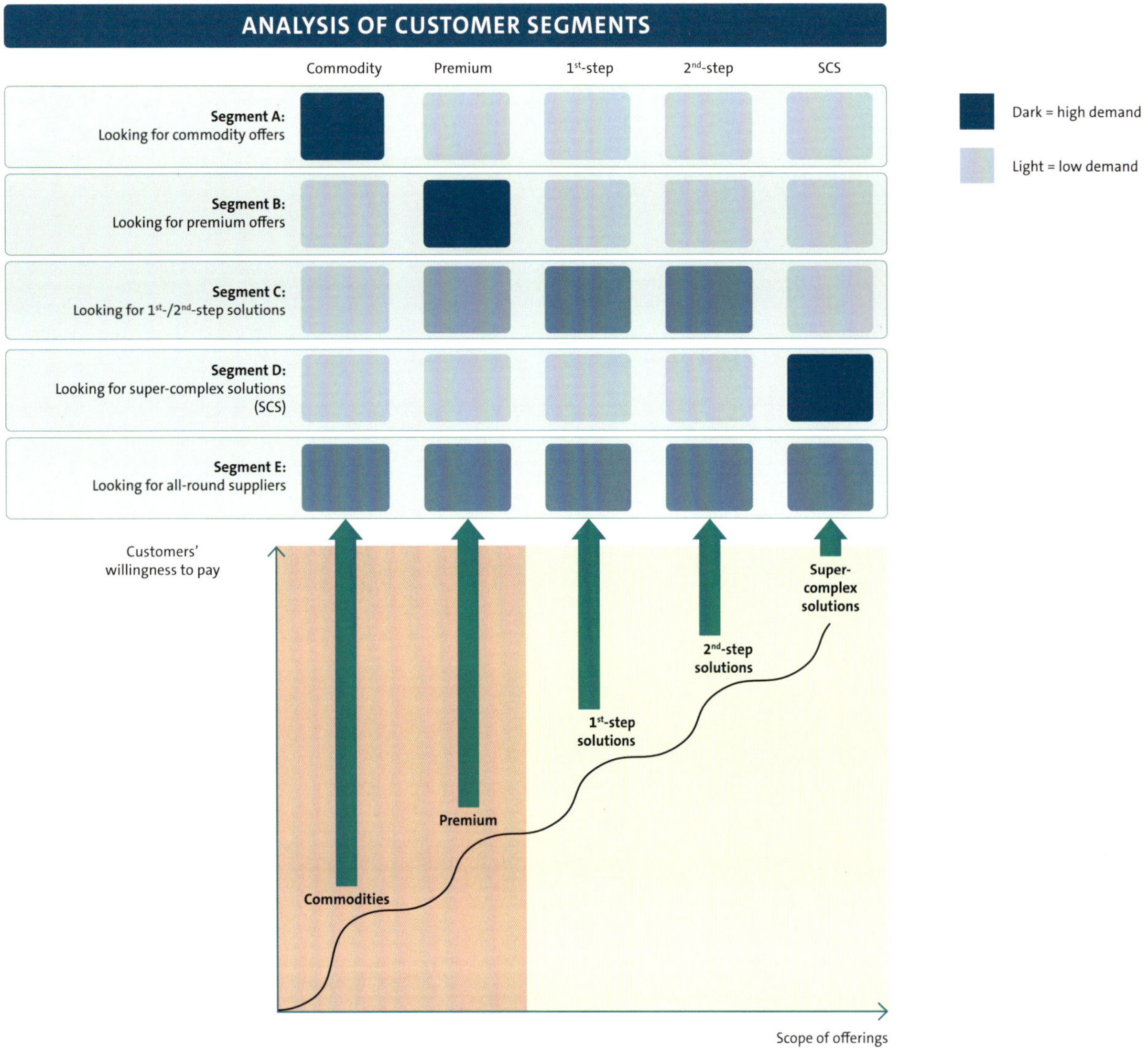

- **Customer segment A ("Looking for commodity offers").** This customer segment is shown on the first line. It only has one dark box on the far left. Demand in this customer segment is therefore focused clearly on commodities. Customers in this segment expect certain minimum standards to be met. If this is ensured, all that they are otherwise interested in is the price.
- **Customer segment B ("Looking for premium offers").** Customers in this segment are also clearly focused. Instead of the price, high quality is at the very top of their needs hierarchy. Accordingly, demand in this segment is focused on premium products. A higher price is justified for customers in this segment if the higher product quality results in noticeably better economy for their own processes.
- **Customer segment C ("Looking for 1st- and 2nd-step solutions").** This customer segment includes customers that have recognized the high benefit of 1st- and 2nd-step solutions for themselves and now use these to meet larger parts of their demand.

  These customers are often too small for individually tailored super-complex solutions (SCS). Smaller customers are lacking the infrastructure to be able to purchase and implement such comprehensive offers.

  Many customers in this segment still meet a relevant part of their demand with offers from the classic product sector. When customers in this segment turn to classic products, they are usually focused on premium products that promise increased economy thanks to their quality. This is why the premium box for this segment is also darker in the figure. Conversely, this means that premium customers offer the most fertile ground for the transition to 1st- and 2nd-step solutions.

  It is certainly conceivable that this segment could split into two segments in highly developed markets in the future ("Looking for 1st-step solutions" and "Looking for 2nd-step solutions"). Often, they cannot yet be differentiated in the current, primarily young solution markets.
- **Customer segment D ("Looking for super-complex solutions").** These customers are reliant on individual, complex solutions that precisely match their special needs. Medium and large customers are often found here, many of which operate with highly specialized business models and are therefore hardly able to implement standard solutions.
- **Customer segment E ("Looking for all-round suppliers").** The major key accounts that want to be fully equipped by their suppliers are usually found here. These companies' purchasing departments expect low prices in the product sector and, at the same time, access to all solution steps, up to and including highly individualized super-complex solutions.

STRATEGY

For an understanding of the market, it is important for you to assess the distribution of customers and the shares of sales for your own market. You should also forecast the shift over the next three years. The following work template helps you to do this (see Figure 53).

Fig. 53 Work template for analyzing the customer segments

						DEVELOPMENT OF CUSTOMER SEGMENTS					
	Commodity	Premium	1st-step	2nd-step	SCS	% number of customers today	% number of customers in 3 years	% of sales volume today	% of sales volume in 3 years	% of profit today	% of profit in 3 years
Segment A: Looking for commodity offers						___ %	___ %	___ %	___ %	___ %	___ %
Segment B: Looking for premium offers						___ %	___ %	___ %	___ %	___ %	___ %
Segment C: Looking for 1st-/2nd-step solutions						___ %	___ %	___ %	___ %	___ %	___ %
Segment D: Looking for SCS						___ %	___ %	___ %	___ %	___ %	___ %
Segment E: Looking for all-round suppliers						___ %	___ %	___ %	___ %	___ %	___ %

The figures are specific to each sub-market and must therefore be determined individually.

A few basic tendencies can be determined for the customer segments with regard to a) the size of the customers, b) the number of customers, c) the sales volume per segment, and d) "customer readiness for solutions."

a) Size of the customers per segment. Small accounts usually have hardly any demand for super-complex solutions, which are not profitable for them, as their setup costs are too high. Small accounts are therefore distributed primarily over product-related segments A and B, whereas the very large customers are usually found in segment E, as described above. Individualized solutions are particularly important here because specific standards have often evolved in major companies. However, many key accounts additionally demand the entire spectrum of offer options from their suppliers in this case. This leads to high complexity costs for the suppliers when serving these key accounts. These costs have to be taken into consideration in the pricing for the business model portfolio.

b) Number of customers per segment. There is also a trend here. The majority of customers are usually found in segment A, followed by segment B, and then segment C a lot further down the line. Comparatively few customers are usually distributed over segments D and E.

c) Sales volume and profit shares per customer segment. The sales volume and profit shares per customer segment are usually distributed differently to the number of customers. A comparison of the number of customers per business model step and their sales volume and profit shares is therefore extremely relevant.

An interesting distribution can be observed for many B2B companies: Fewer than 20% of customers are responsible for at least 80% of the sales volume and profits. When the actual full costs to serve are taken into consideration, losses are often generated in the small account business. This is highly important for an assessment of customer potentials (see Chapter 12) and for deriving sales structures (see Chapter 13).

d) "Customer readiness for solutions." Above all, 2nd-step solutions and super-complex solutions necessitate a high level of professional structures, such as IT structures on the customer side. Even if the demand is there and the customer could potentially derive a benefit from the offer, it is not yet clear whether the customer has the necessary professional structures. Since this "customer readiness" is not one of the elements of traditional customer analysis, it is often overlooked. That is understandable in the classic product business, since no special prerequisites are necessary on the part of customers, but such an omission can prove disastrous in solution selling.

Observe customer segment shift. It is important for sales to know how customer segments shift over time because a shift in customer segments calls fundamental *sales structures* (Do we need personal selling at all for segments A and B, or is a functioning online channel sufficient?) and *sales methods* (What form of selling do the business model steps require in the solution sector?) into question. We also have to rethink our *customer prioritization* (see Chapter 12). It should be oriented to the profit contribution that each segment can make.

B. Analysis of the competitor segments: Where is the strongest competition to be anticipated?

The next step involves analyzing the competitor segments. The "*strategic group analysis*" is a suitable instrument for this. It is a classic tool for determining where the most important supplier groups are located in a market. However, the analysis is no longer familiar to many people today, although it is one of the established tools of classic industrial goods marketing and has been used for numerous markets such as airlines, elevators, automotive, construction, food, and steel. My colleague Professor Mario Rese, who passed away far too early, impressively demonstrated this in his dissertation on the elevator market.[16]

In principle, the strategic group analysis is nothing other than what takes place during customer segmentation. This instrument is used when there are so many suppliers operating in a market that a precise analysis of each individual competitor is difficult. It forms the suppliers into groups according to specific dimensions. The strategic groups encompass suppliers that are similar to one another but differ from the other groups.

As in the case of customer segmentation, supplier grouping is concerned with how the groups can be subdivided. It has been determined that the strategic supplier groups in many markets can be mapped along the customer segments. The supplier groups therefore reflect the customer segments.

In markets with five business model steps, this results in five competitor segments or "strategic groups" (see Figure 54). They pursue positioning strategies that are each coordinated optimally to one of the customer segments. There are therefore usually suppliers with a clear focus on commodities (group A) and suppliers with a clear focus on premium products (group B). Conversely, group C includes suppliers that still sell products but are focused on selling 1st- and 2nd-step solutions. Companies that concentrate exclusively on selling super-complex solutions (group D) as well as all-round suppliers that operate on each level (group E) can also be observed.

It is now important to determine the supplier segment in which particularly strong competitors are encountered. This analysis should include not only the current situation, but also the prognosis for the coming years.

[16] Rese (1993), *Technische Normen und Wettbewerbsstrategie (Technical Standards and Competitive Strategy)*, dissertation, Berlin: Beuth.

Fig. 54 Comparison of customer and competitor segments

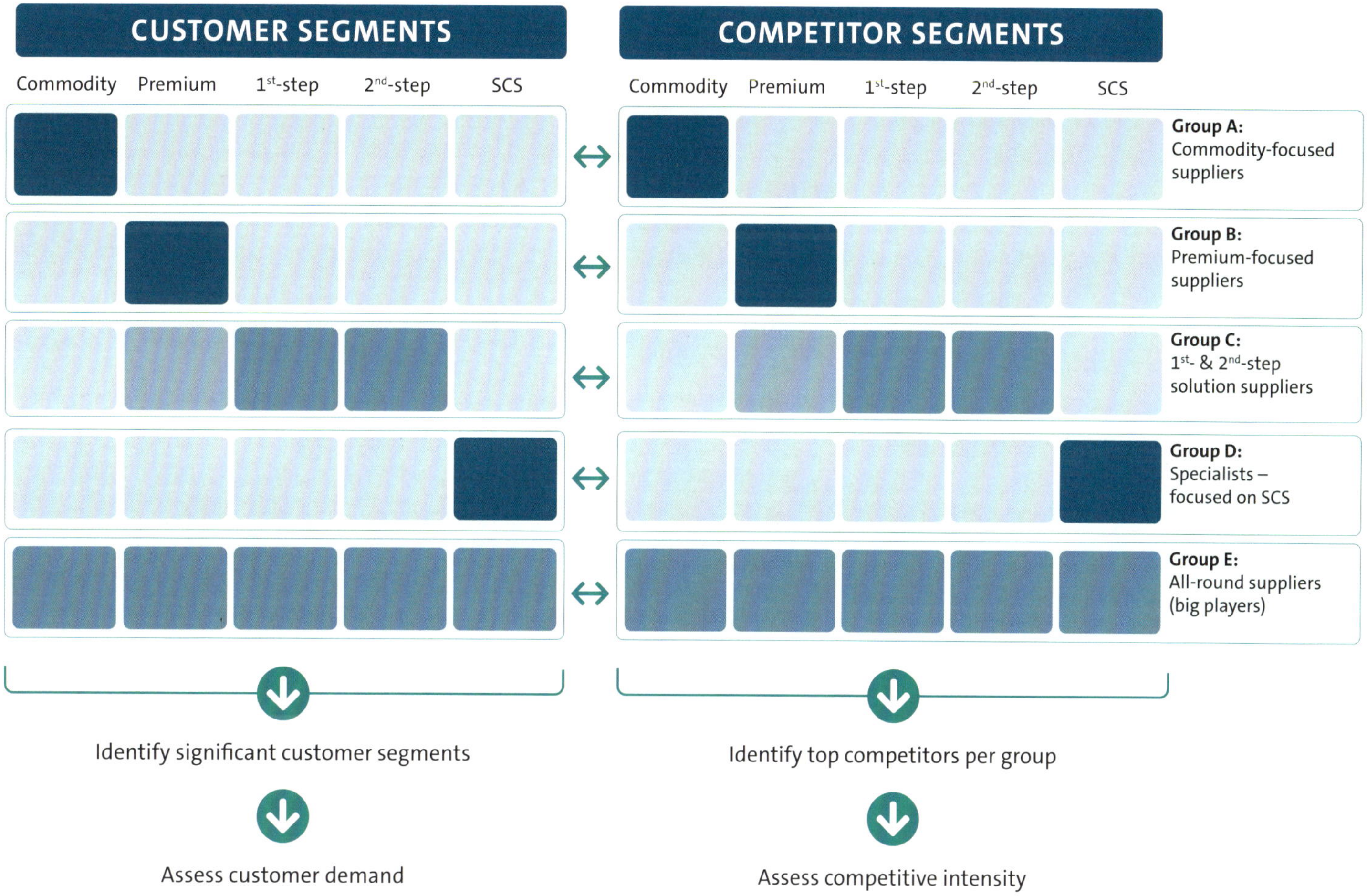

C. Derivation: Where are the biggest market opportunities located for your business models?

In the final step, the business models with the biggest market opportunities are identified. To do this, customer demand and the competitive intensity are summed up per business model.

Once both the customer segments and the strongest competitors per segment have been identified within a business model step, this information can be used to derive the future competitive advantage potentials. As shown in Figure 55, customer demand is placed in relation to the competitive intensity for each business model step.

Fig. 55 Assessing market opportunities per business model step

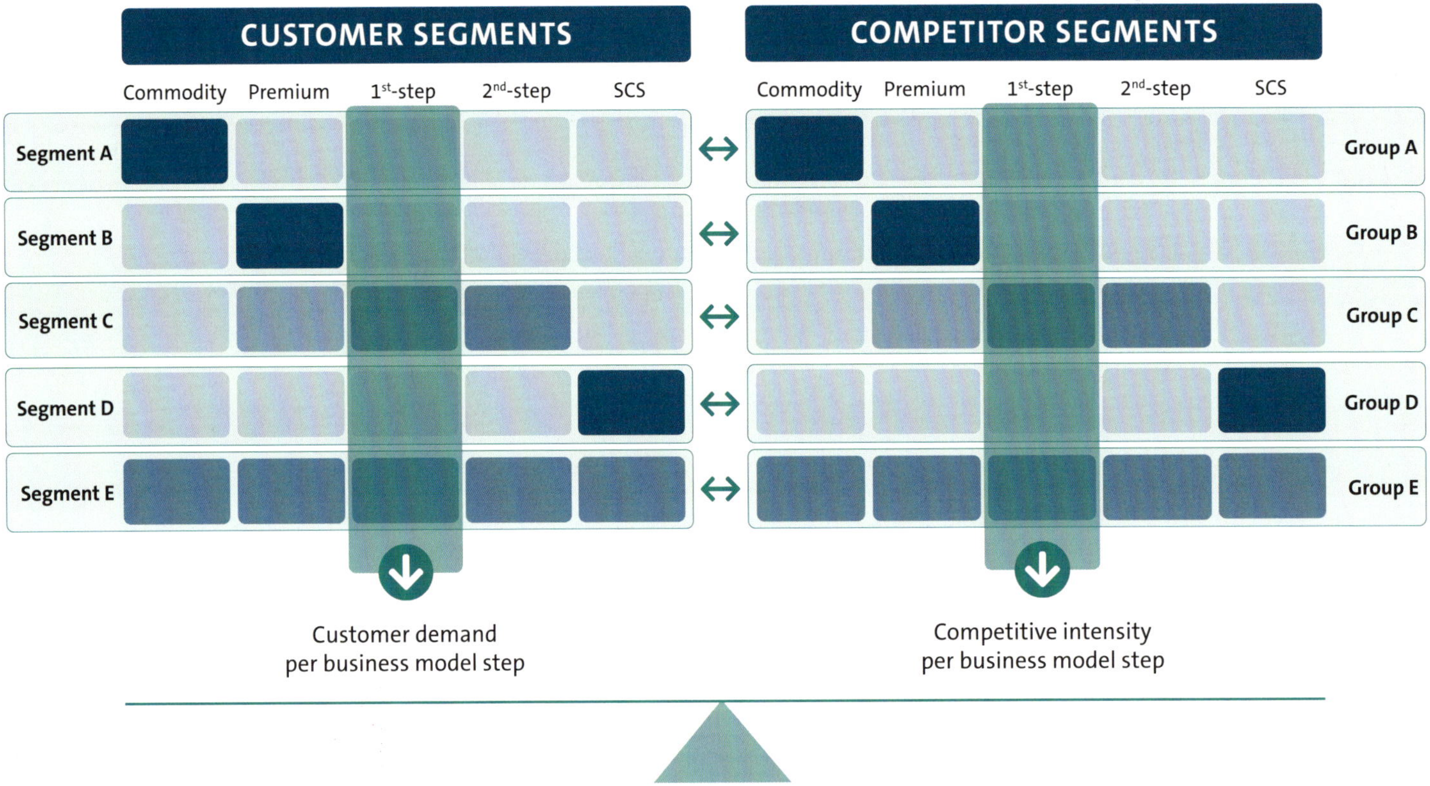

The following table contains a checklist for determining the extent to which you already have sufficient know-how regarding your business models.

Checklist – business model know-how				
1. Are you familiar with the most important business model steps for your market and the most important offers of your competitors on each step?	Yes	☐	No	☐
2. Do you know precisely which business model step you are on with each of your offers?	Yes	☐	No	☐
3. Can you classify the most important characteristics of your solution business models using the "SNOOFY" factors?	Yes	☐	No	☐
4. Do you regularly analyze which product/service offers in your market still lie in the lack of transparency range today but will presumably be digitally comparable in three years (keyword *"comparability limit"*)?	Yes	☐	No	☐
5. Have you clearly defined the extent to which you are pursuing an "Atriplex strategy" (price/cost leadership) vs. a solution business model strategy?	Yes	☐	No	☐
6. Have you analyzed the most important benefit drivers in the customer value chains for your target customer segments?	Yes	☐	No	☐
7. If you select three customers at random, can the responsible account manager tell you precisely what the most important benefit aspects of your solution business model are for these customers?	Yes	☐	No	☐
8. Is the account manager familiar with the risks due to which these customers would advise other customers against purchasing the solution?	Yes	☐	No	☐
9. Are you familiar with the characteristics in the offerings of your business models that engender the greatest willingness of customers to pay?	Yes	☐	No	☐
10. Have you conducted a scenario analysis (best, middle, worst cases) concerning your long-term cost development in your solution business models?	Yes	☐	No	☐
11. Have you checked whether your current sales setup meets the requirements of the planned business model step?	Yes	☐	No	☐
12. Are you familiar with the customer segments per business model step in your market?	Yes	☐	No	☐
13. Are you familiar with the competitor segments per business model step in your market?	Yes	☐	No	☐

In the third part of the book, I would like to present practical examples of 1st-step solutions, 2nd-step solutions, and super-complex solutions. To do this, I have created 18 business model profiles for companies in various industries. They are consciously structured so that the reference to the six SNOOFY factors is clear in each case.

WHY SALESTEGY?

STRATEGY – FINDING THE RIGHT BUSINESS MODELS

PRACTICAL EXAMPLES

SALES – IMPLEMENTING NEW BUSINESS MODELS IN SALES

Chapter 7

18 SNOOFY business model profiles

- Business model profiles for 1st-step solutions
- Business model profiles for 2nd-step solutions
- Business model profiles for super-complex solutions

Message of this chapter

Six of the 18 business model profiles are concerned with 1st-step solutions, six with 2nd-step solutions, and six with super-complex solutions.

What should you have learned by the end of the chapter?

- Solution business models are found in all industries.
- Each solution business model can be structured using the SNOOFY factors.
- A total SNOOFY score can be determined for each business model.

The following table provides an overview of the distribution of the business model profiles with reference to the business model step, the industry, and the size of the companies.

No.	Company	Industry	Company size		
			Large (more than $5 billion)	Medium (up to $5 billion)	Small/startup
		1st-step solutions			
1	Kelvion	Industrial goods & mechanical engineering		x	
2	jetlite	Technology & software			x
3	PHYSEC	Technology & software			x
4	Kieback&Peter	Energy & supply		x	
5	Continental	Automotive	x		
6	Uponor	Energy & supply		x*	
		2nd-step solutions			
7	Trox	Energy & supply		x*	
8	ALD	Industrial goods & mechanical engineering		x*	
9	CWS	Others		x	
10	Wilo	Industrial goods & mechanical engineering		x*	
11	Hilti	Construction, industrial goods & mechanical engineering	x		
12	WISAG	Industrial goods & mechanical engineering		x*	
		Super-complex solutions			
13	Kühne+Nagel	Transport & logistics	x		
14	GEA	Industrial goods & mechanical engineering	x*		
15	SAP	Technology & software	x		
16	Schneider Electric	Conglomerate	x		
17	IBM	Technology & software	x		
18	Siemens	Conglomerate	x		

* "Hidden champion" = Hidden champions are international market leaders in their industry, but comparatively less known to the broader public. The term was coined by Hermann Simon.

Business model profiles for 1st-step solutions

Business model profile no. 1: Kelvion – solution for increasing the performance and extending the life cycle of industrial heat exchangers

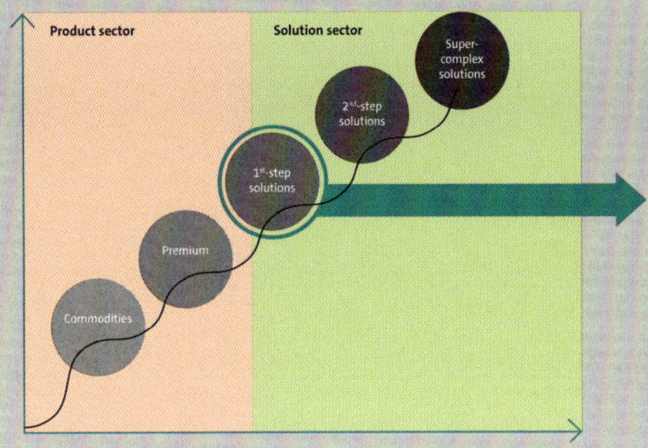

Company: Kelvion Holding

Supplier of: Industrial heat exchangers

Sales: $1.1 billion

Industry: Industrial goods & mechanical engineering

What is marketed?	Increasing the performance and extending the life cycle of industrial heat exchangers by taking over various process steps (design, production, commissioning, maintenance, servicing)
To whom is it marketed?	Energy industry, oil and gas industry, chemicals industry, food and drinks industry, computer centers as well as building and refrigeration technology
For which added benefit are customers willing to pay?	*"With our broad product and service portfolio, we enable heat exchangers to develop their maximum benefit at customers and therefore allow customers to structure their processes more efficiently. We work intensively together with our customers and optimally adapt our solution to specific customer needs. Our heat exchanger technologies and the accompanying service offer also ensure that energy costs can be reduced, possible downtime costs can be avoided, and the life cycle is extended."* Marcel Aßmann – Vice President Commercial Marketing & Sales Strategy at Kelvion

Kelvion is a traditional German company with its headquarters in the Ruhr region. It has evolved to become one of the world's leading experts on heat exchangers and has more than 67 sites and partners around the globe. Since 1920, the company has been producing heat exchangers of all kinds for industrial application areas, for example in the oil, gas, and chemicals industries. A heat exchanger ensures that heat is transferred from one substance to another. In this process, one of the substances dispenses the heat and the other receives it. The task of industrial heat exchangers is to sensibly use heat that would otherwise be lost for other processes, for example. The heat exchange processes each take place spatially separately (e.g., via heat-conducting metal chambers or pipes) so that no mixing takes place and each substance remains in its circuit. *Kelvion* sells both individual heat exchanger components and complete heat exchangers in the premium segment.

Kelvion is also active in the service business on a global scale. The objective is to cover all of the process steps surrounding the life cycle of a heat exchanger (commissioning and on-site service; replacement parts and replacement part solutions; monitoring, advice, and training as well as upgrades and exchange).

The benefit of this solution arises due to the fact that each process step is optimized by *Kelvion* and the heat exchangers achieve their maximum efficiency and performance *(**Y**ield customer value-in-use)*. Preventive maintenance measures that are digitally integrated ensure that the heat exchangers always perform to the best of their capabilities. Among other aspects, this enables significant savings in primary energy consumption and results in the extension of the operating time. Another positive effect of this predictive maintenance from *Kelvion* is that heat exchanger downtimes – and therefore also the related financial losses – are avoided. Pricing is focused on the long-term increase in the heat exchangers' performance *(**O**utcome-based pricing)*.

Kelvion increases the efficiency of the heat exchangers by supporting all necessary process steps in its solution *(**N**urture customer processes)*. In the first step, the specific requirements of the heat exchangers are analyzed on site together with the customer. After professional consulting, *Kelvion* develops the heat exchangers, already during this early project phase, so that the customer can optimally adapt them to its specific application areas. After production, *Kelvion* technicians assemble and install the heat exchangers.

However, *Kelvion's* cooperation with its customers does not come to an end even after commissioning. The comprehensive service offer helps to ensure that all services still come from a single source in bundled form *(**S**ervitization)*. To do this, *Kelvion* offers regular inspections for the heat exchangers. Among other things, service technicians check whether leaks occur or whether corrosion protection has to be applied in this process. If necessary, they modernize the heat exchangers and exchange individual parts. As described above, these services lead to the fact that *Kelvion* guarantees maximum performance, extends the operating time of the heat exchangers, and ensures that the customer can avoid financial losses due to downtimes.

The intensive cooperation with customers during the design phase and the extensive services offered by this solution contribute to an intensive and long-term customer relationship *(**F**ar-reaching relationship)*.

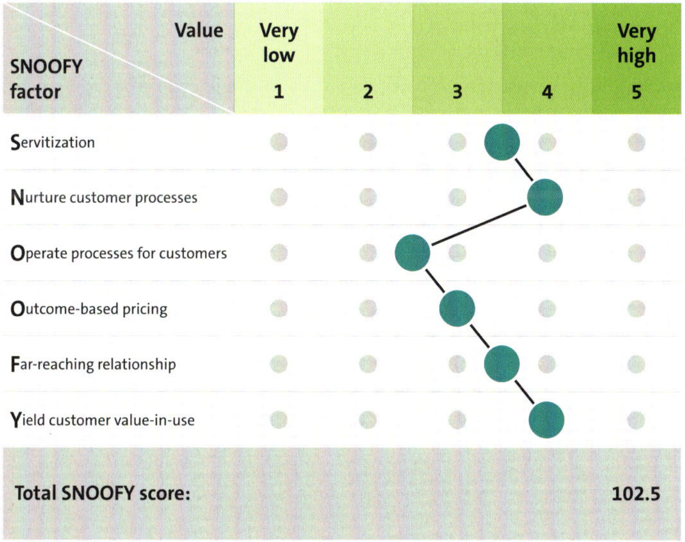

Business model profile no. 2: jetlite – light control solution for improving human chronobiology

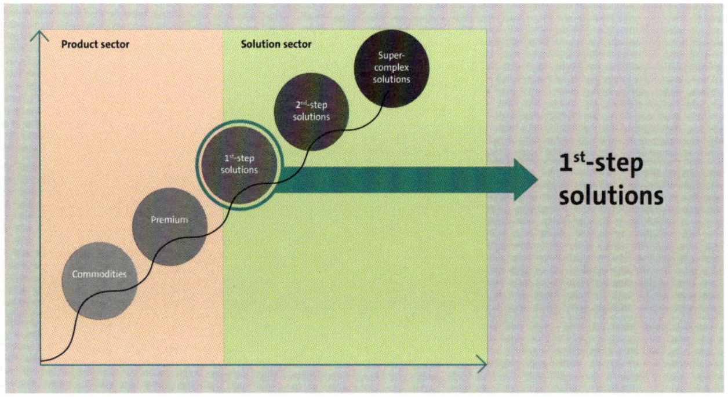

Company: jetlite

Supplier of: Light control software

Sales: ~ $1 million

Industry: Technology & software

What is marketed?	Software-based light control solution that adapts the light environment to the biological rhythm of humans and optimizes it
To whom is it marketed?	Tradespeople in the industrial sector (offices), healthcare, and the hotel industry (hospitality)
For which added benefit are customers willing to pay?	*"Our software-supported light control solution enables optimal adaptation to the biological rhythm of humans on the basis of algorithms. We thereby help our customers to improve the concentration and well-being of their employees and customers."* Dr. Felix Brüggemann – Managing Partner at jetlite

jetlite is a startup that develops light control solutions for optimizing human well-being and concentration. To do this, *jetlite* uses research findings from the field of human-centric lighting to offer light concepts based on chronobiological analyses and learning algorithms. *jetlite* originally developed its solutions for aircraft cabins in order to reduce jet lag and other negative physical symptoms on long-haul flights. Meanwhile, the company is especially active in the industrial sector, healthcare, and in the hotel industry.

The light control solution adapts the light design to the human body clock and thereby improves well-being, concentration, and health. Based on algorithms, a light concept individually adapted to the customers' situation is developed and supported with additional services.

The "benefit during use" can be explained by using a hospital as an example *(**Y**ield customer value-in-use)*. For instance, the light control solution optimizes the natural circadian rhythm in patients and reduces the negative physical symptoms caused by the rhythm in a hospital, which usually differs significantly from the natural human rhythm. Thanks to special chronobiologically active lighting controlled by a *jetlite* algorithm, the patients' recovery time is reduced from an average of four to three days. Above all, the price of the solution portfolio offered by *jetlite* is oriented to the lighting modules used by the customers. Added to this are model calculations for optimizing the customers' productivity *(**O**utcome-based pricing)*.

With this solution, *jetlite* helps its customers to improve their services by improving the employees' working environment and increasing the well-being of guests or patients *(**N**urture customer processes)*. In order to continuously improve the solution, *jetlite* provides analysis tools that operate on the basis of artificial intelligence.

jetlite designs and implements the light control solution individually for every customer. *jetlite* additionally remains in contact with the customer companies even after the purchase has been completed by means of various after-sales services *(**S**ervitization)*. For instance, the company continuously advises on how to use the solution optimally. In this process, *jetlite* helps its customers to conduct studies with which the optimal effectiveness of the solution can be validated externally. *jetlite* also helps its customers to market the innovative lighting solution.

jetlite is in constant and close contact with its customers in order to optimally adapt the light control solution to the customers' respective situation *(**F**ar-reaching relationship)*. Due to its long-term cooperations, *jetlite* succeeds in continuously evolving its own algorithm and in generating an added value for both parties.

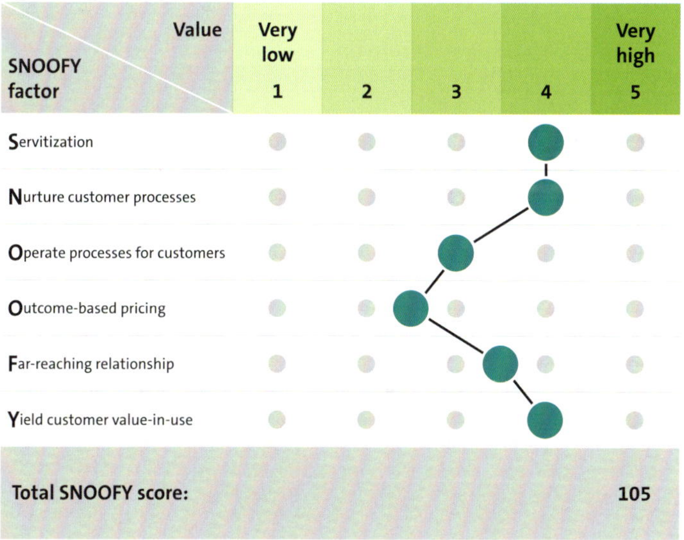

Business model profile no. 3: PHYSEC – innovative security solution in the Internet of Things (IoT)

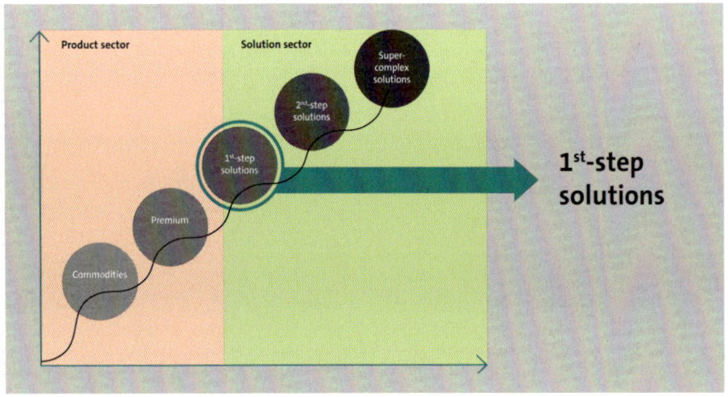

Company: PHYSEC

Supplier of: Security solutions for applications in the IoT

Sales: > $1 million

Industry: Technology & software

What is marketed?	Security solutions for applications in the IoT
To whom is it marketed?	Companies with critical infrastructures, primarily from the utility industry, for example power grid operators
For which added benefit are customers willing to pay?	*"Our customers' willingness to pay is particularly due to the fact that we offer maximum security for their IoT hardware. Because of that, we protect our customers from cyber attacks, among other things, and secure their data using cutting-edge encryption technologies. We stand by our customers even after implementation and continuously optimize the solution to ensure constant protection."* Dr. Heiko Koepke – CFO at PHYSEC

PHYSEC is a startup that is specialized in IoT security solutions, for which it has already won multiple prizes (including one as one of the top 10 startups in Germany). The company's name stands for *physical-layer security* and describes a technology that secures corporate activities in the IoT. *PHYSEC* refers to this as *security for things*. *PHYSEC's* customers are primarily utility industry companies, including major power grid operators, whose infrastructures are critical.

With its solution, *PHYSEC* provides its customers with an IoT platform *(IoTree)* that serves as the interface between the customer's physical and digital worlds. On this platform, all of the customer company's hardware data is recorded and made digitally usable. The platform can be implemented in the public cloud, private cloud, or on-site cloud.

PHYSEC implements hardware modules in the customers' terminals. These automatically generate a digital fingerprint for all machines and processes, enabling them to be monitored via the IoT platform. In the event of a physical attack or technical manipulation or if damage occurs, the system sounds the alarm. *IoTree* therefore optimizes the customers' security system *(**Y**ield customer value-in-use)*. The security solution's pricing is oriented to the number of terminals connected and the scope of the consulting services provided by *PHYSEC (**O**utcome-based pricing)*.

In this business model, the service content is high *(**S**ervitization)* because *PHYSEC* offers various services around the platform. *PHYSEC* employees assist the customer already during the initial test phase so that customers can test the solution in full. *PHYSEC* also offers workshops and further training that help the customers' employees to assess the security status and identify improvement potentials.

Starting with its initial analyses, *PHYSEC* supports customer processes *(**N**urture customer processes)*, for example, by simulating an attack on the customers' systems by cyber criminals to detect vulnerable areas in the devices. *PHYSEC* then analyzes the customer's system architecture to determine optimization potentials.

Throughout the entire cooperation, *PHYSEC* remains in continuous dialog with its customers and works closely with them to continuously evolve their security systems *(**F**ar-reaching relationship)*.

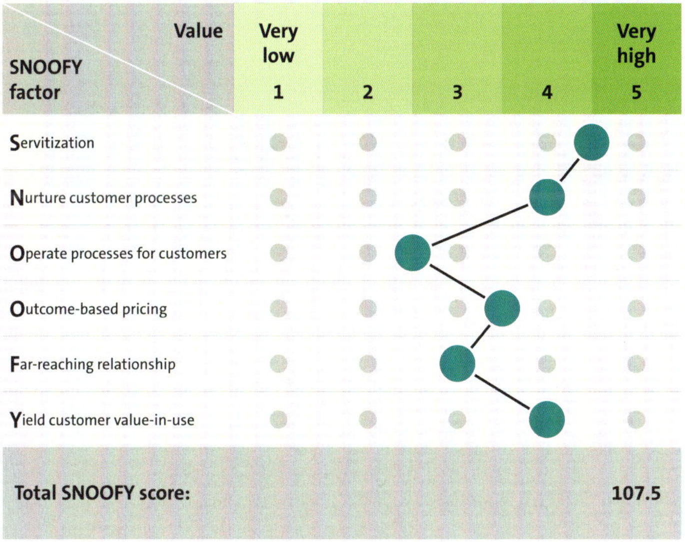

Business model profile no. 4: Kieback&Peter – *RIEcon Smart Building* system solution for digital real estate management

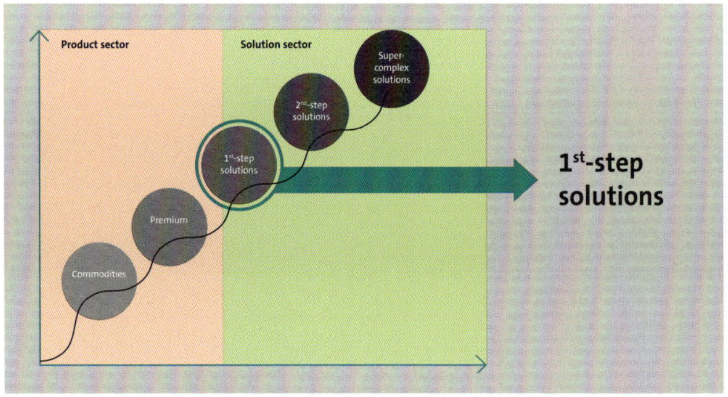

Company: Kieback&Peter

Supplier of: Building automation technology

Sales: $234 million

Industry: Energy & supply

What is marketed?	Digital Smart Building solution for the efficiency-enhancing management of buildings and residential areas
To whom is it marketed?	Real estate administrators, real estate construction companies, realty companies
For which added benefit are customers willing to pay?	*"Our variably scalable Smart Building system solution is used both to automatically manage buildings and residential areas and also to determine a baseline for the energy consumption and emissions output of the building stock of the real estate rental industry. Permanent monitoring of energy flows significantly improves the energy and CO_2 balance. We achieve enormous time-saving potentials for our customers by digitalizing complex energy-efficiency administrative processes for them."* Christian Gronwald – Segment Manager Residential&Living Germany, Kieback&Peter

With solutions for more than 30,000 apartments, *Kieback&Peter* offers an extensive wealth of experience in networking residential areas and public facilities and in managing energy flows for decentralized energy supply in keeping with the energy revolution.

The *RIEcon Smart Building* system solution is the digital communications platform between the operator of energy centers and the energy consumers in the residential areas. Thanks to this platform, customers can clearly see their energy consumption and CO_2 emissions at any time, leading to the omission of laborious administrative processes such as reading meters. In turn, this leads to effective time management and reduces labor costs. In addition, the individual components regulate resource-efficient energy input using self-learning artificial intelligence. These efficiency-enhancing systems reduce heating energy consumption by up to 20% *(**Y**ield customer value-in-use)*. The software also offers online system visualizations to reveal possible anomalies in energy consumption. Laborious manual maintenance processes are replaced by this solution, as the location of malfunctions in the customer's energy supply can be seen in digital form *(**N**urture customer processes)*. The pricing for this solution varies depending on the number of apartments and buildings that are networked, and it gives consideration to provision and implementation costs *(**O**utcome-based pricing)*.

Kieback&Peter individually tailors the digital automation solution to customers' needs depending on the size of the real estate and the number of rented properties. The company also undertakes the implementation of the corresponding automation technology. To this end, each apartment/rental unit is equipped with a *RIEcon AppartmentManager* system, each building with a *RIEcon BuildingManager* system, and the entire area with a *RIEcon QuartiersManager* system. These are networked to one another by *Kieback&Peter* service technicians. *Kieback&Peter* also integrates the solution in the customers' digital systems *(**O**perate processes for customers)*.

Kieback&Peter additionally offers its customers workshops and training courses to guarantee that they use the software efficiently and profitably *(**S**ervitization)*.

Kieback&Peter works closely together with its customers during the planning and implementation process. The individual scalability of the solution also fosters further cooperation because the customers' other residential buildings can also be easily integrated in the existing system *(**F**ar-reaching relationship)*.

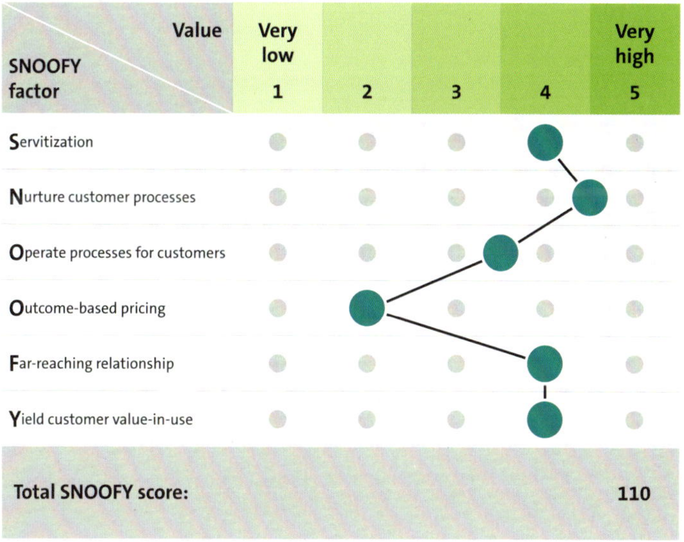

Business model profile no. 5: Continental – digital and sustainable tire management with *Conti360° Fleet Solutions*

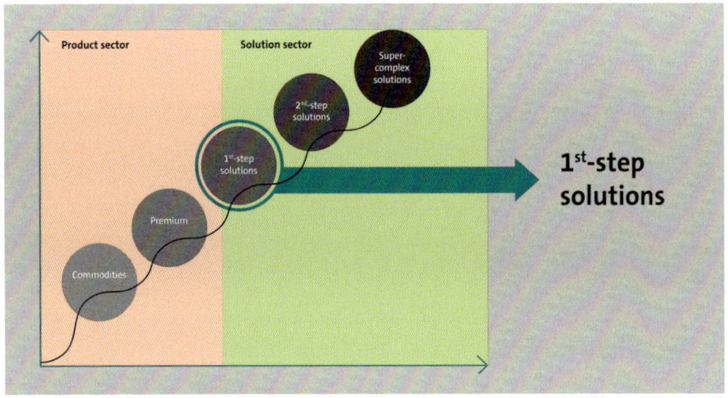

Company: Continental

Supplier of: Tire and vehicle services

Sales: $44.2 billion

Industry: Automotive

What is marketed?	Integrated fleet solution: From tires (passenger cars, vans, trucks), tire service, and digital tire solutions up to and including sustainability through retreaded tire solutions
To whom is it marketed?	All companies with vehicle fleets: For example, transport companies (freight transport), rental and leasing, authorities, passenger transport, construction industry
For which added benefit are customers willing to pay?	*"Continental regards itself not only as a supplier of premium tires, but, particularly in the fleet business, as a professional partner to its customers for jointly developing important solutions for future topics. As one of the world's biggest suppliers of electronics, sensor systems, and software in the mobility industry, the company wants to become the market leader for service-based, digital solutions by 2030. The integration of digital tire monitoring solutions into the Conti360° solutions contracts is the logical next step in accompanying transport industry customers on their road to digitalization."* Annika Lorenz – Head of Fleet Solutions Germany at Continental

Continental is one of the global market leaders in tire sales and characterized by its offers surrounding truck, passenger car, and bicycle tires. *Continental* employs more than 230,000 staff at more than 550 locations in 58 countries worldwide.

Continental is consistently continuing its evolution from a tire and service supplier to an integrated solution supplier. In the future, fleet solutions will be an overall package: The *Conti360° solutions contracts* encompass both digital solutions such as *ContiConnect*, *ContiPressureCheck*, and *ContiLifeCycle* with the retreaded tire products *ContiRe*, *ContiTread*, and carcass management.

The customer benefit of *Conti360° Fleet Solutions* is that customer needs can be individually structured and covered through modular contract options. In this way, *Continental* offers the transport industry a professional solution for managing fleets efficiently, subject to sustainability aspects: with the use of fuel-saving tires, resource-friendly tire retreading, and digital tire monitoring.

In sum, *Continental's* offer is aimed at ensuring a longer service life for tires and minimizing the risk of failure. Truck fuel consumption can also be significantly reduced *(**Y**ield customer value-in-use)*. Among other methods, *Continental* optimizes vehicle fleet performance by providing customers with the tire monitoring and analysis data in digital form for their fleet managers and drivers *(ContiConnect)*. The prognoses based on this data increase the safety of the tires because the necessity of exchanging or repairing the tires is pointed out in good time. Both processes are carried out by *Continental* partner workshops if necessary. Flat tires can be avoided in this way. *Continental* offers two cost systems for *Conti360° Fleet Solutions*. Customers can choose between a service contract that is invoiced exactly according to the services used or a mileage-based contract. This is extensively dependent on performance KPIs (e.g., miles, fleet size), and thus the benefit that the customer obtains over time thanks to the solution *(**O**utcome-based pricing)*.

The service content of the business model is high *(**S**ervitization)*. At the beginning of the partnership, *Continental* draws up a concept for the customers. In this process, the company coordinates all services individually with its customers. These include comprehensive carcass handling, 24/7 customer service that is available year-round, and reliable and fast roadside assistance.

Supporting the customers' processes is also important *(**N**urture customer processes)*. In this case, *Continental* identifies potential savings at the very start of the partnership on the basis of collected customer data (e.g., wheel number and mileage, sensor data). This data is processed for the customers and made available in the form of digital requirement forecasts. All tire values are made available to the fleet managers within a web portal *(ContiConnect)*. *Continental* contacts them pro-actively via various communication channels. The fleet drivers are also provided with tire pressure and temperature values in real time and are warned immediately in the event of deviations *(ContiPressureCheck)*.

The SNOOFY factor ***F**ar-reaching relationship* is extremely pronounced in *Conti360° Fleet Solutions*, as the services are guaranteed continuously over a number of years, and tire management is continually being improved through the latest technologies. The supplier and customer work closely together to achieve this and develop a relationship that is maintained over several years.

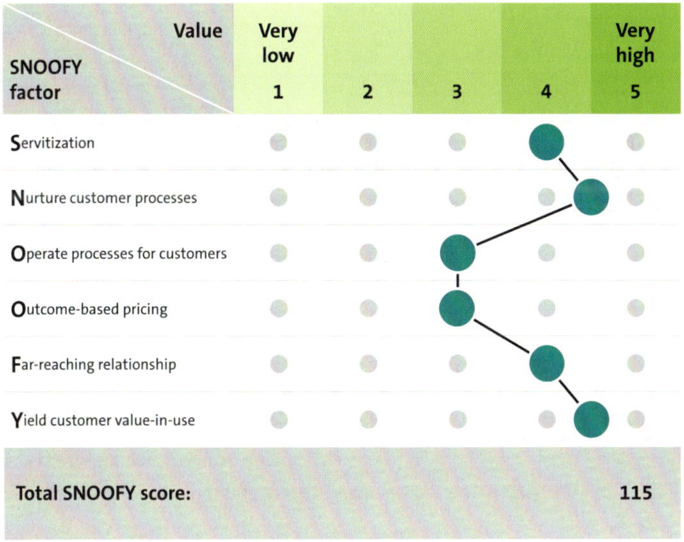

Business model profile no. 6: Uponor – planning solution for building supply systems

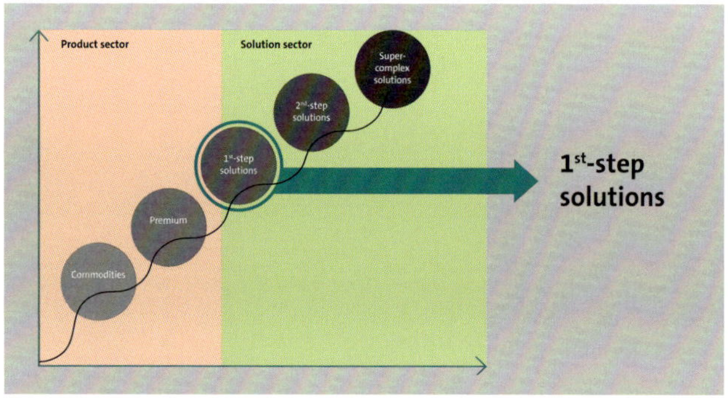

Company: Uponor

Supplier of: Heating, cooling, and drinking water systems

Sales: $1.3 billion

Industry: Energy & supply

What is marketed?	Planning of energy-efficient, sustainable supply systems (heating, cooling, and drinking water systems) in residential and commercial real estate construction
To whom is it marketed?	Residential and commercial construction, particularly construction clients, architects, planners, and trade contractors
For which added benefit are customers willing to pay?	*"We offer digital planning and implementation of building supply systems. We support our customers intensively before and during construction. This also applies to the application of the solution after completion. We enable significant savings in water and material consumption and therefore increased profitability."* Dr. Karsten Hoppe – President Building Europe & International, Member of the Executive Committee at Uponor

Finnish stock corporation *Uponor* has evolved from a small Scandinavian manufacturer of plastic pipes to a hidden champion in heating, cooling, and drinking water systems. In the product business, the company previously sold premium system components for plastic and composite pipes that are used in drinking water installations as well as floor heating and cooling systems.

It now has an additional business model. With the aid of Building Information Modeling (BIM) and the subsequent installation of the heating, cooling, and drinking water systems, *Uponor* is able to plan sustainable, energy-efficient building technology. The company's customers include residential and commercial real estate construction companies, construction clients, architects, planners, and trade contractors.

By guaranteeing efficient building supply with its solution business model, *Uponor* improves the profitability of its customers and acts in an ecologically sustainable manner *(**Y**ield customer value-in-use)*. Ultimately, there are potentials for saving significant amounts of water and energy. Since the efficient building technology means that hardly any maintenance or servicing costs are incurred, *Uponor* customers' financial outlays for supplying the buildings are reduced. Using the BIM data, *Uponor* makes it possible to plan the entire building supply system efficiently and to minimize errors in installing the systems. *Uponor's* solution also increases the attractiveness of the buildings: The constant temperature and humidity levels ensure a pleasant interior climate. *Uponor's* pricing is oriented to the procurement costs of the heating, cooling, and drinking water systems. However, this is supplemented by an individual calculation that takes the forecast increase in economy in operating the building into consideration *(**O**utcome-based pricing)*.

In this business model, *Uponor* concentrates on supporting its customers' processes *(**N**urture customer processes)*. When planning their buildings, for instance, customers can use *Uponor* guidelines that enable technically matured and sustainable heating technology and drinking water installations. *Uponor* also supports its customers' planning with digital planning software. This can be used as a mobile app or a desktop version, for example for 3D supply system installation planning or for planning panel heating and drinking water modules. Once the construction work is completed, *Uponor* supports its customers by using digital software tools to detect supply anomalies and to clearly analyze and optimize internal consumption.

Uponor offers customers additional services within its solution *(**S**ervitization)*. These include enabling the construction client to locate and contact specialist tradespeople directly using a search portal provided by *Uponor* as well as training courses to ensure that the *Uponor* planning software is operated properly.

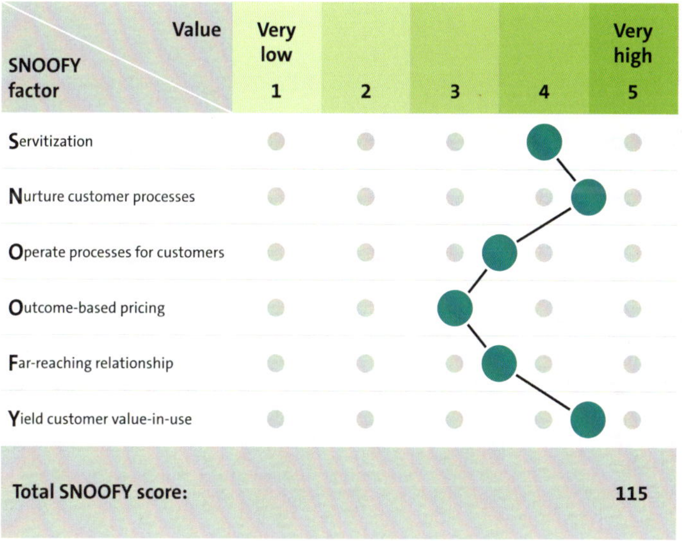

Business model profiles for 2nd-step solutions

Business model profile no. 7: TROX – air handling system solution

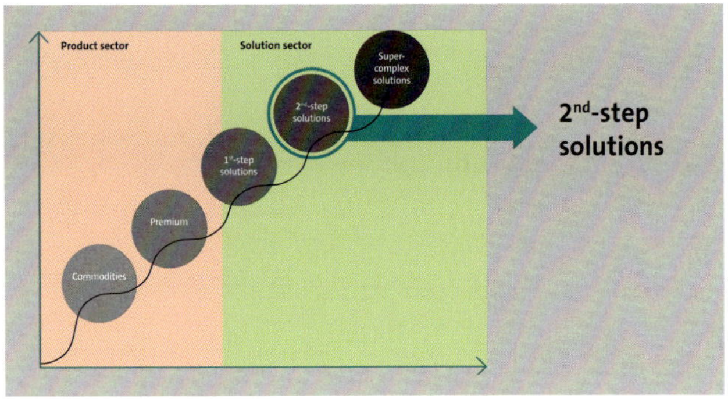

Company: TROX

Supplier of: Air handling systems

Sales: $604 million

Industry: Energy & supply

What is marketed?	Energy-efficient operation of air handling systems (e.g., air conditioning, ventilation, fire protection, and smoke control systems)
To whom is it marketed?	Owners of, for example, hotel, hospital, and laboratory buildings
For which added benefit are customers willing to pay?	*"We offer an air handling system solution characterized by energy efficiency and safety. We preventively ensure that the air handling systems function optimally and look after them throughout their entire service lives."* Martin Alofs – Head of Product Management Air Handling Units at TROX

Family-owned company *TROX* is a hidden champion and describes its business as "The art of handling air." It manufactures products for air handling systems (e.g., air conditioning, ventilation, fire protection, and smoke control systems). These include filter elements, control units, and air diffusers. The concept for the respective systems is usually drawn up by external planning offices, subsequently coordinated with the operator, and procured, installed, and commissioned by the system manufacturer. Many other trades, such as steel construction, shell construction, and electricians, are also involved in the planning phase.

In the business model for its air handling system solution, *TROX* goes beyond the product supplier level because the company creates a complete air handling solution from a single source in order to facilitate the complex coordination of the involved trades for its customers *(**N**urture customer processes)*. The core of this is the digital and thus intelligent networking of all components and devices in order to achieve more efficient overall control of the air conditioning and ventilation systems in total. This is why work starts for *TROX* in the planning phase. During this, the integration of the air handling systems in the architectural planning is prepared. *TROX* also coordinates the installation of the systems and offers performance tests for the installed components. The customers of these systems often come from sensitive sectors such as pharmaceuticals industry laboratories.

Thanks to its air handling system solution, *TROX* shortens the industry's usual construction times *(**Y**ield customer value-in-use)*. *TROX* achieves this by taking over the installation of the air handling system, thus ruling out coordination problems between the various trades. The solution also enables energy savings and high system reliability during system operation. Digital system simulations, with which the systems' energy consumption can be analyzed and subsequently optimized, are carried out to achieve this. *TROX* takes account of the faster construction time and the efficiency of the systems in its pricing *(**O**utcome-based pricing)*.

TROX therefore takes over essential processes for its customers *(**O**perate processes for customers)*. For instance, the air handling technology in laboratories and hospitals has to comply with special safety standards and meet the specific requirements of the customers. To uphold this commitment to quality, all air handling system planning and development is carried out by *TROX's* internal engineering department.

Preventive remote diagnosis and regular maintenance intervals ensure a high service content *(**S**ervitization)*. *TROX* also has service teams ready and waiting to assist its customers at any time in the event of problems or questions. The customer relationship extends over a longer period of time than in *TROX's* classic product business and is especially intensive in the planning and installation phase *(**F**ar-reaching relationship)*.

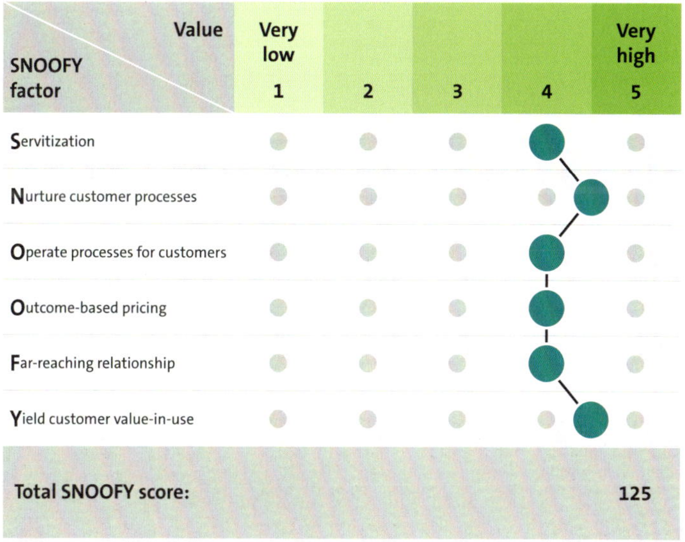

Business model profile no. 8: ALD EXPERT for the digital transformation of production processes in the areas of metallurgy and heat treatment

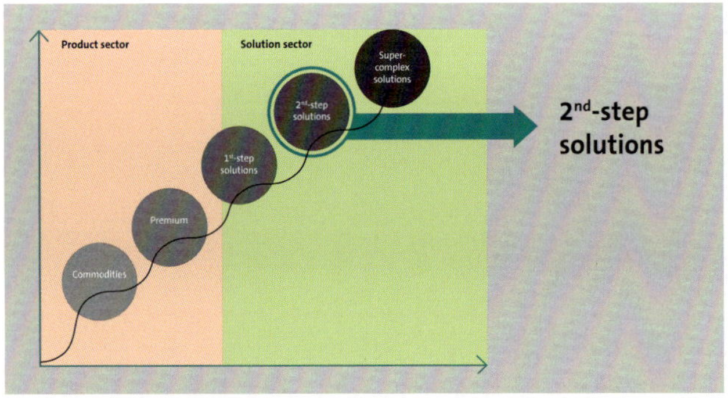

Company: ALD Vacuum Technologies

Supplier of: Vacuum production systems

Sales: $129 million

Industry: Industrial goods & mechanical engineering

What is marketed?	Intelligent digitalization and automation solutions for the industrial sector, focusing on production systems in the areas of vacuum metallurgy and heat treatment
To whom is it marketed?	Primarily ALD's existing customers in the automotive, aerospace, medical technology, special metals, and chemicals industries as well as other industries in the industrial sector
For which added benefit are customers willing to pay?	*"Our customers benefit from our experience and in-depth knowledge in process and plant engineering in combination with our solution platform (ALD EXPERT). This means that, today, we are in the position to share our experience and findings in dominant future topics such as digital transformation, energy and resource management, knowledge management, and cyber security."* Harald Kirsch – Head of Automation & Industrial IT at ALD Vacuum Technologies

ALD Vacuum Technologies GmbH is a hidden champion whose roots extend back to 1916, when the company succeeded in smelting chromium-nickel alloys under vacuum conditions for the first time. The company is now the world's leading manufacturer of vacuum systems in the fields of smelting technology, metallurgy, and heat treatment. Among other things, the various vacuum systems and furnaces are used to manufacture solar cells, for applying the special coating on turbine blades, and for the heat treatment of high-precision engine parts and tools. Accordingly, these systems are used in the automotive industry, in manufacturing aircraft and electricity turbines, and in the energy industry.

At *ALD*, the concept of the *ALD EXPERT* solution business model arose with the aid of digitalization. In this business model, *ALD* develops and implements an individual concept for the digital transformation of vacuum production systems together with its customers. The objective is to transform, automate, and optimize the existing process and production flows using innovative digital technologies.

Digital transformation of the production systems by *ALD EXPERT* provides an extensive benefit, particularly due to increases in efficiency *(**Y**ield customer value-in-use)*. This solution digitally maps all production and manufacturing processes, and it enables the customer to obtain information about the current status of production at any time using a mobile terminal. In addition, process parameter tracking and monitoring (PROCESS DATA VISUALIZATION and BATCH MANAGEMENT) ensure that the production batches offer constant quality and that unscheduled failures and disruptions are avoided. Specially programmed monitoring algorithms and mobile 3D system visualizations help the customers to detect unusual operating states early on and to introduce the necessary measures to eliminate them (ANOMALY DETECTION). This innovative maintenance process (Maintenance 4.0) significantly reduces production downtimes. The use of digital technologies therefore increases system effectiveness and helps to optimize the customers' production processes. This benefit is reflected in the pricing. The price of the solution is also oriented to the value that the solution generates for the customer *(**O**utcome-based pricing)*.

ALD develops a solution concept together with the customers and undertakes the integration of the innovative technologies in the vacuum production systems *(**O**perate processes for customers)*. In the design phase, the customer's situation is first analyzed precisely using process and data analytics, and all resources related to the system process are recorded (RESOURCE MANAGEMENT). The energy consumption of the system and the entire production plant is also recorded to determine energy-saving potentials (ENERGY MANAGEMENT). The systems' relevant data is then linked together so that gap-free and object-related documentation of all processes can take place over the entire useful life of the systems. Later on, these data links help to determine irregularities during the production process and to monitor the system status from mobile terminals. This leads to the improvement of the production flows *(**N**urture customer processes)*.

Due to the intensive cooperation of *ALD* and the customers on site, an individual digitalization concept is developed within *ALD EXPERT*. Once the concept has been integrated, *ALD* remains its customers' first point of contact, particularly when replacement parts or repairs are necessary. Interaction with customers before and after the integration of *ALD EXPERT* is therefore intensive *(**F**ar-reaching relationship)*.

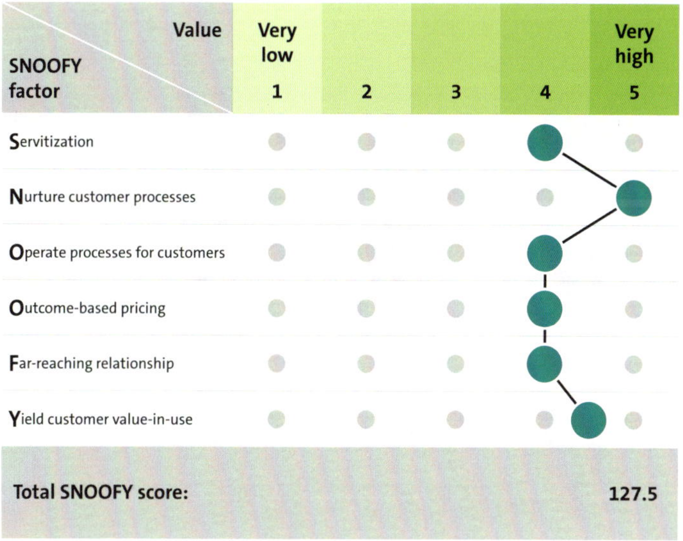

Business model profile no. 9: CWS – *loovio* washroom hygiene solution

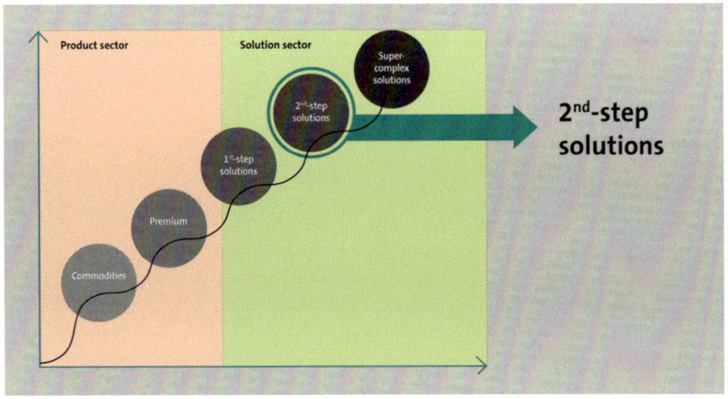

Company: CWS International

Supplier of: Hygiene equipment and workwear

Sales: $1.3 billion

Industry: Others

What is marketed?	Provision and operation of washroom systems in commercial real estate and public buildings
To whom is it marketed?	Operators of commercial real estate (e.g., hotel operators, service station operators) and public buildings
For which added benefit are customers willing to pay?	*"With loovio, we offer all services surrounding the efficient operation of highly frequented, public washrooms. Our offer covers everything from construction planning to the installation of our innovative and sustainable sanitary facilities. This includes the delivery of hygiene products, operational cleaning and care of the washrooms, and the maintenance of the components. Our full service provides our customers with everything from a single source and enables them to achieve significant cost savings and focus even better on their core activities."* Heike Seltman – Group Director Sales at CWS

CWS (formerly *CWS-boco*) is a hidden champion that offers its customers the hire and the sale of hygiene equipment, floor mats, laundry services, fire prevention systems, and workwear, etc., with more than 11,000 employees worldwide.

In its *loovio* solution business model, *CWS* undertakes all services to ensure the efficient operation of washroom systems in commercial real estate (e.g., hotels, service stations) and public buildings. These include the individual compilation of the washroom system, digital analysis of the consumable materials used, and maintenance of the washroom components.

loovio offers customers multiple advantages. Since they no longer have to cooperate with a variety of different suppliers for consumable materials, their overhead and material costs are reduced. By selecting a smart wash solution, they can also reduce water consumption by up to 90% and soap consumption by up to 60% *(Yield customer value-in-use)*. Costs are additionally lowered by using the digital analysis tools offered by *CWS* to analyze and optimize the consumption data, fill levels, and availability of the washroom components. For chargeable washroom systems, such as those at service stations, *CWS* offers voucher systems that enable additional revenue. The price of the *loovio* solution is negotiated individually with customers. It is oriented to the long-term benefit that the customers achieve through the operation of the washroom systems (conditional ***Outcome-based pricing***).

In the *loovio* solution, *CWS* takes over the processes of installing the washroom systems, maintenance, and the replacement of consumable materials *(Operate processes for customers)*. For instance, customers are supplied with products such as towel and soap dispensers for their washrooms, the washrooms are maintained, and cleaning staff are provided. The consumption of washroom materials can also be monitored and analyzed using digital technology. One familiar example of taking over value-adding customer processes is the *loovio* solution for service stations. In this, service station operators are provided with a complete solution offer for their washrooms from *CWS*.

By operating the washroom systems, *CWS* remains in constant contact with its customers *(Far-reaching relationship)*. The corresponding customer employees and *CWS* are in regular dialog and work continuously on improving the washroom solution. The intensity of the relationship is also reflected in the length of the partnerships; at an average of 10 years, it is comparatively long for the sanitary and washroom industry.

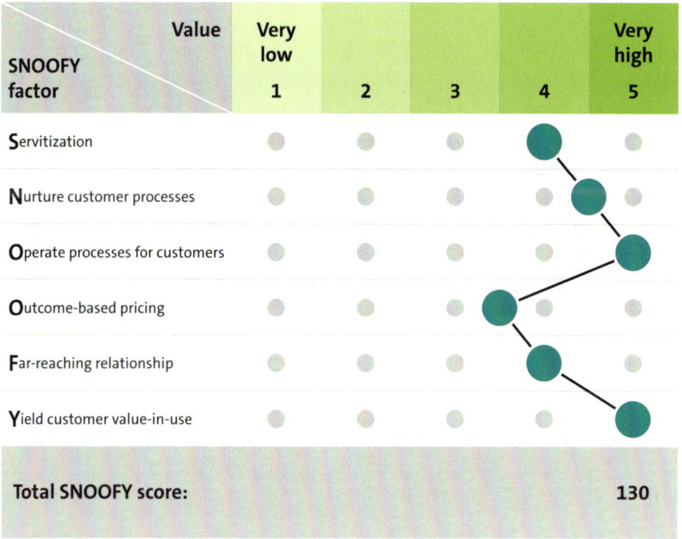

Business model profile no. 10: Wilo – pump system solution for large buildings

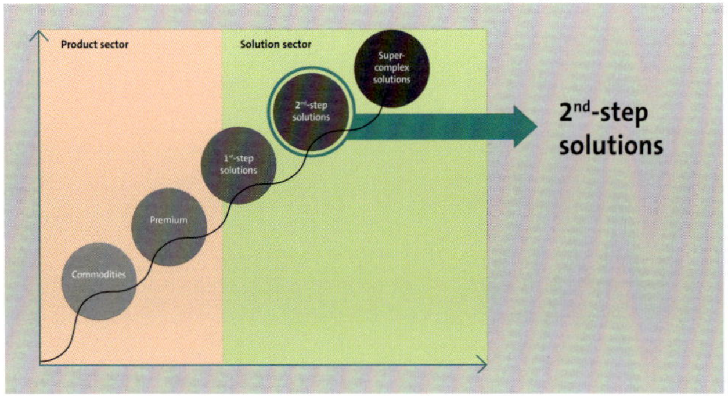

Company: Wilo

Supplier of: Pumps and pump systems

Sales: $1.8 billion

Industry: Industrial goods & mechanical engineering

What is marketed?	Optimized water supply and climate control in buildings
To whom is it marketed?	Operators of large buildings (e.g., hospitals, office complexes, and hotels)
For which added benefit are customers willing to pay?	*"With our solution, we guarantee our customers the right water pressure anywhere in the building. Thanks to this, we enable economically optimized water supply and climate control. And all with maximum reliability because our predictive maintenance system alerts us even before the customers notice anything."* Marc Stiebing – Senior Vice President Sales Region Mature Markets at Wilo

Wilo SE is a hidden champion from the classic product business. *Wilo* offers both commodity and premium pumps for heating, refrigeration, and air conditioning, water supply as well as wastewater disposal and cleaning. *Wilo* is now investing extensively in solution business models. These include connective pump solutions such as the *Wilo-Assistent app*, which helps specialist tradesmen to plan, install, and commission pumps, for instance (see p. 83).

One of the most complex solutions from *Wilo* is a pump system solution for large buildings (e.g., hospitals, office complexes, hotels) that ensures optimal and reliable water supply and climate control. To achieve this, intelligent pump systems are installed at various locations in the buildings and digitally networked.

The customer benefit of this pump system solution is that it reduces energy costs by up to 80%. *Wilo* assumes responsibility for optimal climate control and water supply, which is guaranteed around the clock *(**Y**ield customer value-in-use)*. *Wilo* can remotely monitor and control the customer's pump systems using a predictive maintenance system. This is made possible by the intelligent, digital networking of the pumps and pump systems, and it enables *Wilo* to react early on and exchange potentially defective components so that the customers do not suffer any loss of performance. *Wilo* can also convert the pump systems easily and modernize interface modules. This enables uncomplicated responses if the customer changes its software systems in the future. This minimizes pump downtimes and protects the customers' investment in the long term. The extensive assumption of responsibility by *Wilo* within this solution increasingly shifts the customers' results into the focus of the pricing *(**O**utcome-based pricing)*.

In implementing this solution, *Wilo* is initially focused on improving customer processes *(**N**urture customer processes)*. Together with its customers, the company determines which pumps are needed in which locations in the buildings. This necessitates intensive cooperation with the customers, in which multiple decision-makers such as plumbers, planners, and the end-users of the buildings are involved *(**F**ar-reaching relationship)*. Following installation, *Wilo* takes over the operation of the pump systems. Throughout their entire service lives, the pump systems are looked after by a specialized team of *Wilo* experts *(**O**perate processes for customers)*.

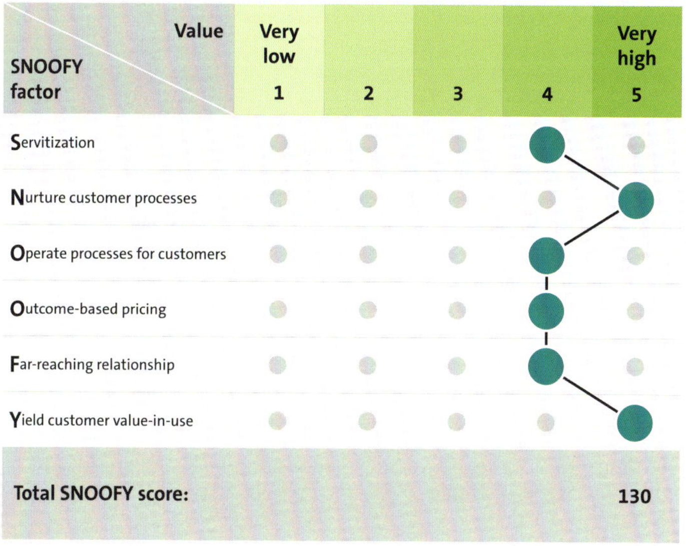

Business model profile no. 11: Hilti – *Building Information Modeling (BIM)* solution

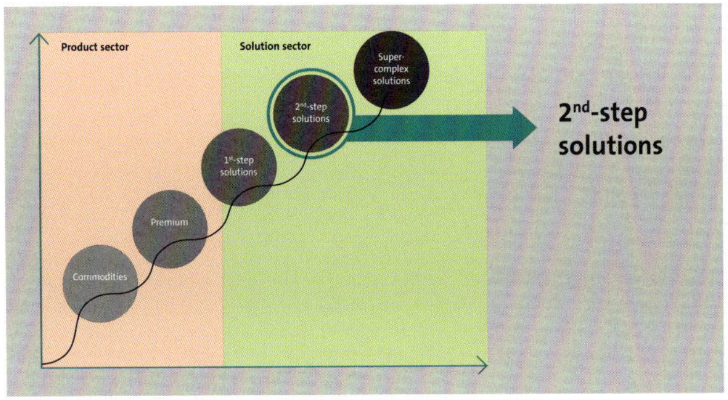

Company: Hilti

Supplier of: Tools, fastening materials

Sales: $5.6 billion

Industry: Construction; industrial goods & mechanical engineering

What is marketed?	Detailed, model-based planning service for modular fastening systems and passive fire protection that enables numerous use cases for increasing productivity during the construction process
To whom is it marketed?	General contractors, technical building equipment companies as well as building owners from the construction industry (e.g., structural and civil engineering, steel, metal, and facade construction) that obtain an added value from increased workflow efficiency
For which added benefit are customers willing to pay?	*"As a manufacturer with direct contact to all relevant construction partners, we understand which BIM use cases actually lead to productivity increases for the customer. For the BIM with Hilti solution, we integrate our expertise in modular fastening and passive fire protection with services along the entire value chain. This reduces coordination effort in the technical building equipment environment as well as installation times by up to 50% compared to the usual industry values."* Daniel Gmeiner – Head of BIM Competence Center, Hilti Group

Hilti AG is an internationally leading tool manufacturer with around 30,000 employees worldwide. Its high-performance premium products (e.g., cordless devices, industrial vacuum cleaners, devices for direct installation) are used in more than 120 countries.

Early on, the company recognized that the construction industry will change radically in the future due to ongoing digitalization. *Hilti* subsequently developed a partnership-based, integrative solution based on the BIM methodology that currently represents the company's most complex solution business model. The *BIM with Hilti solution* is used to optimize all aspects of modular fastening technology and passive fire protection, which are critical to the course of construction in the early planning phase by means of digital designs and modeling. The digital twin created in this way enables an information-based increase in productivity in all subsequent process steps.

Design optimizations and prefabrications up to and including the *BIM-2-Field solution* lead to significantly faster construction and material cost savings of up to 50% *(**Y**ield customer value-in-use)*. The aim of the integrative planning for fastening the technical building equipment is to detect and avoid possible collisions at an early stage. The standardized solutions are prefabricated in modular form, leading to "as-designed" quality and minimizing installation times. For positioning the system solutions or for automated drilling, an intelligent *BIM-2-Field* process enables the efficient and exact implementation of the simulated construction model. In this, ultra-modern measurement technology is used to precisely replicate the digital planning at the construction site so that the work procedures can be implemented clearly and efficiently in the right place at the right time. As a result, ad-hoc decisions on site are minimized and subsequent problems and malfunctions stemming from hasty decisions by construction managers are avoided. Buildings supported by *Hilti's BIM solution* during construction are less prone to malfunctions in any case in the long term. The price of the *Hilti BIM solution* is oriented to the benefit derived by the customers *(**O**utcome-based pricing)*.

In the first step, *Hilti* specifies all construction requirements together with the customers and then integrates them down to the very last detail in the planning software. *Hilti* additionally advises its customers on selecting the optimal products in order to lay the foundation for efficient construction work during this early project phase *(**N**urture customer processes)*.

Hilti then takes over the planning process and develops the entire construction project based on the previously defined project requirements *(**O**perate processes for customers)*. The *BIM*-based modeling offers the basis for efficient construction work. *Hilti* also prepares detailed workshops and overview drawings that provide the necessary clarity for successful construction work on the construction site. *Hilti* additionally coordinates the various trades for its customers with punctual logistics and transparent cycle sequences *(**S**ervitization)*. Pre-conversion, pre-cutting, and pre-assembly reduce the assembly effort and the material costs. Detailed bills of materials and quantity calculations produced by *Hilti* ensure cost control and save customers time and money.

This solution intervenes deeply into the customers' operational planning and construction processes in addition to supporting and improving them. The intensive cooperation during concept creation and the integration of the construction requirements into the software lead to a very pronounced SNOOFY factor ***F**ar-reaching relationship*.

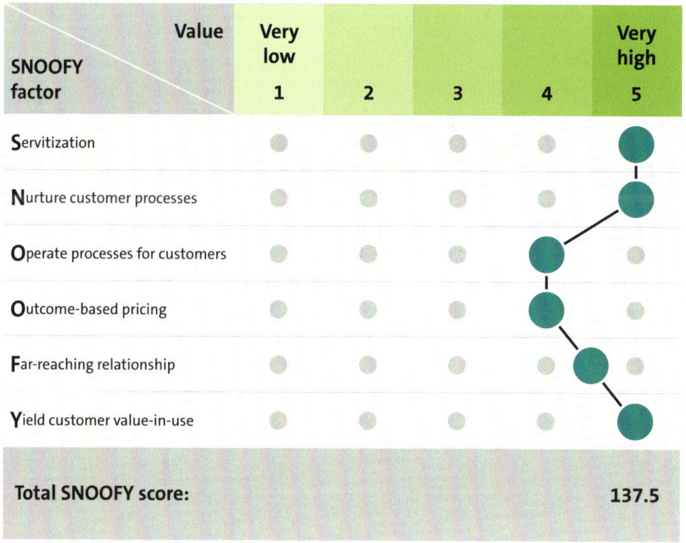

Business model profile no. 12: WISAG – operator model for industrial customer facilities

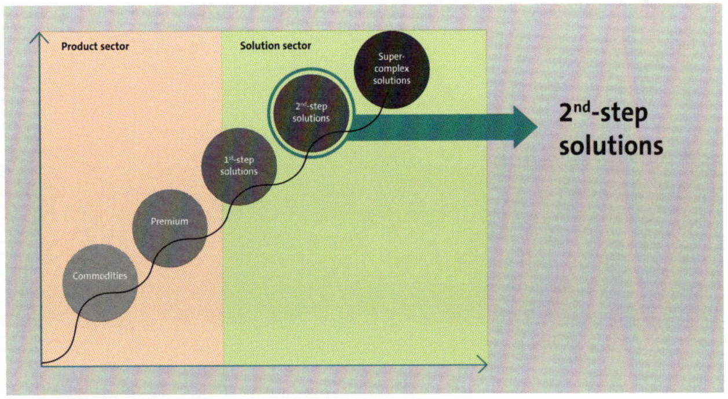

Company: WISAG Holding

Supplier of: Facility, industry, and aviation services

Sales: $2.7 billion

Industry: Industrial goods & mechanical engineering

What is marketed?	Among other things, the establishment and operation of industrial customer facilities (e.g., production and distribution facilities, technical building equipment)
To whom is it marketed?	Production companies in the industrial environment (e.g., utility companies, chemicals industry)
For which added benefit are customers willing to pay?	*"We set up and operate individual customer facilities according to the wishes of our customers and guarantee the optimal implementation of their processes. This minimizes the risk of facility failures, as a result of which high cost savings and efficiency increases are possible."* Stephan Hovestadt – Head of Refrigeration and Air Conditioning / Maintenance Business Unit / North-West

Founded in 1965, *WISAG* is a hidden champion in the field of industrial services and has around 50,000 employees. Its subsidiary *WISAG Building and Industry Service* is specialized in the planning, installation, and maintenance of technical building equipment, including refrigeration, air conditioning, and air handling systems in the industrial environment.

In its most complex solution, *WISAG Building and Industry Service* independently sets up and operates entire customer facilities (e.g., air conditioning and air handling systems). With this solution, *WISAG* offers its customers significantly more cost-effective processes while guaranteeing the optimal functionality of the facilities at the same time. In this, *WISAG* takes on long-term responsibility for its customers' facilities and guarantees their maintenance and operational reliability as well as the minimization of failures.

The *WISAG Building and Industry Service* operator model provides customers with an improved cost structure and increased profitability *(**Y**ield customer value-in-use)*. To do this, *WISAG* optimizes the individual process steps through digital needs analyses and optimal resource management. The pricing for this solution is oriented to the increased profitability that customers achieve by using the solution *(**O**utcome-based pricing)*.

In this operator model, *WISAG* initially works out the precise project requirements together with its customer and then moves on to the detailed planning of the customer facilities. After setting up the facilities, *WISAG's* service technicians guarantee their permanent availability through foresighted maintenance measures *(**N**urture customer processes)*.

The operational processes for operating the facilities are undertaken completely by *WISAG (**O**perate processes for customers)*. This includes the situation that customers have to respond to increased demand by extending their production.

The customer relationships are usually very long. The facilities, which are often technically very complex, also necessitate detailed and intensive coordination between *WISAG* and the customer. A deep understanding of the customer's business model is crucial to the success of the partnership and necessitates a close relationship between *WISAG* and its customers *(**F**ar-reaching relationship)*.

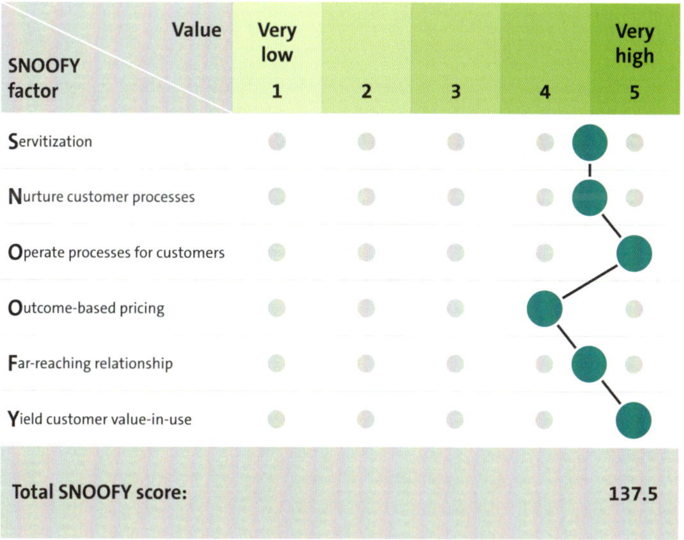

Business model profiles for super-complex solutions

Business model profile no. 13: Kühne+Nagel – *KN International Supply Chain* solution

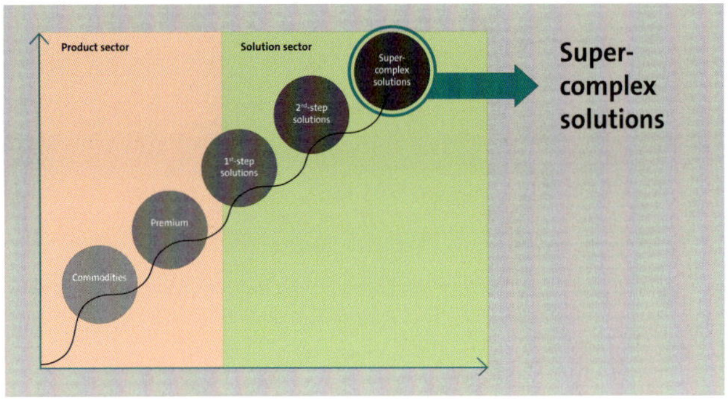

Company: Kühne+Nagel Group

Supplier of: Logistics and freight transport

Sales: $27.5 billion

Industry: Transport & logistics

What is marketed?	Global end-to-end management of supply chains
To whom is it marketed?	Cross-industry customers with logistics requirements
For which added benefit are customers willing to pay?	*"We take on overall responsibility for all of our customers' logistics processes: From ordering components and coordinating suppliers up to and including packaging and, ultimately, safe transfer to our customers' customers. This not only involves outsourcing, but rather digitally optimized co-creation."* Björn Röhlich – Global Head of Sales Training, Kühne+Nagel

The internationally operating logistics company *Kühne+Nagel* is one of the world's leading suppliers over all freight routes (air, water, land) in the logistics and transport industry.

Kühne+Nagel is a full-service provider that is able to link different services, up to and including assuming responsibility for its customers' entire logistical value-creation process, from the end of production to delivery to the customers' customers. This leads to the optimization of customers' supply chains.

The customer benefit is derived from the global assumption of end-to-end responsibility for supply chains by *Kühne+Nagel*. This includes punctual delivery of the undamaged goods to the right location *(**Y**ield customer value-in-use)*. Coordination can be carried out by live monitoring of the transport routes using a GPS tracker for transport protection in the truck, in the aircraft, on the ship, or in the train. In the case of sensitive or cost-intensive products, for instance, a special team can also continuously monitor the temperatures and framework conditions of the goods along the transport route so that sensitive products arrive safely and undamaged at the destination. This monitoring guarantees a transparent supply chain as well as reliable and efficient scheduling for customers. Complex and time-consuming supply chain management is therefore minimized for the customers. With a fully integrated *4PL solution*, the price of this solution can be oriented to the result achieved through digitally optimized co-creation *(**O**utcome-based pricing)*.

Taking over the logistical processes at the customer's site leads to the fact that the SNOOFY factor ***O**perate processes for customers* is extensively pronounced. Such processes include ordering and purchasing goods or their components on behalf of the customers. Transport from the suppliers' locations to the customers' warehouses then takes place. *Kühne+Nagel* operates these warehouses for the customers and undertakes all of the processes that take place within the warehouse. These include the receipt of products, packing or repacking, and shipping to the end customers. A digital solution that records, synchronizes, and optimizes the customers' entire supply chain has been developed for shipping. *Kühne+Nagel* subsequently coordinates the transport routes from the production plants to the point of final sale.

The involvement of customers in concept creation leads to the establishment of customized solutions such as *KN International Supply Chain* or the *4PL Integrated Logistics solution* for complex logistical processes, for example. To achieve this, *Kühne+Nagel* compiles customer-specific project teams and trains them intensively to work closely together with the customers over a long period of time in this business model and to evolve the logistics solution *(**F**ar-reaching relationship)*.

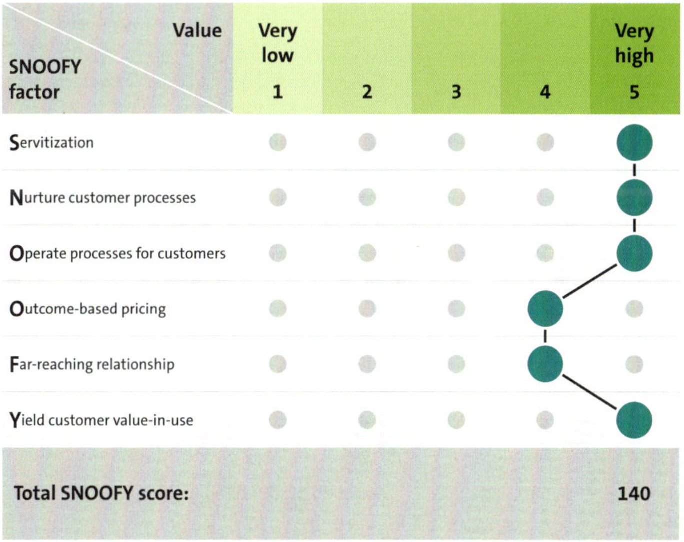

Business model profile no. 14: GEA – complete process solution for downstream food processing

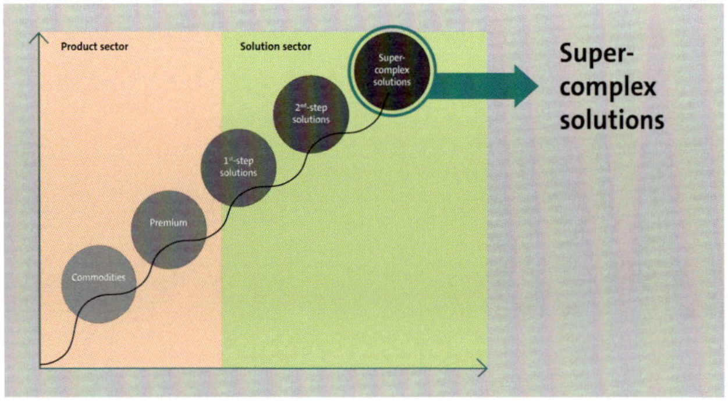

Company: GEA Group

Supplier of: Food processing systems and machines

Sales: $5.4 billion

Industry: Industrial goods & mechanical engineering

What is marketed?	Complete solution for process optimization in downstream food processing (e.g., cooling, preservation, processing, packaging)
To whom is it marketed?	Companies in the food industry and related industries (e.g., food industry, pharmaceuticals industry, chemicals industry, agriculture)
For which added benefit are customers willing to pay?	*"In this complete solution, we undertake digitalized co-creation surrounding their food factories together with our customers. We support, optimize, or take over all of our customers' process steps, from planning and implementing factory conversion/outfitting and the digitalization of processing processes up to and including product preparation, processing, preservation, cooling, and packaging. As a result, we are able to guarantee customers the cost-reducing, efficient downstream processing of their foods."* Tjitze de Wit – CSO of GEA's Liquid & Powder Technology Division

Despite its size, the *GEA Group* is considered to be a hidden champion. The company has established itself worldwide as one of the biggest premium suppliers in the downstream food processing industry and related industries. *GEA* employs around 18,500 staff worldwide and is regarded as a market and technology leader in its business segments. The company offers a diverse range of machines, systems, processing technologies, and components.

In the solution business model analyzed here, *GEA* offers a complete solution for optimizing central downstream food processing steps. In this solution, *GEA* provides all of the technical and digital infrastructure for efficiently preparing, cutting off, processing, preserving, cooling, packaging, and freezing the foods produced by its customers. At the customer's site, the digitally linked process flows of the Industrial Internet of Things (IIoT) ensure the optimization of existing processing processes and the reduction of process and personnel costs as well as guarantee that the factory infrastructure functions.

By guaranteeing the goal-oriented conversion or outfitting of factories, technical process reliability, and the adoption of customers' central processing processes, *GEA* offers its customers, which are generally very large companies in the food industry, the optimal functionality and high efficiency of their secondary food processing *(**Y**ield customer value-in-use)*. The minimization of manual operation steps and failure rates is achieved through the automation and digital networking of processes. For instance, digital analysis tools enable improved product shelf lives during food cooling. In this business model, *GEA* determines the price of the solution by calculating a detailed business case for the customer *(**O**utcome-based pricing)*.

Within this solution offer, *GEA* takes over significant customer processes, including the implementation of the machines and their power supply, the digitalization and analysis of existing processes, and the cooling and climate control of the facilities and plants *(**O**perate processes for customers)*. From the concept design of the food factories to their technical outfitting (e.g., electrical systems and climate control), *GEA* strives to not only provide active support, but also to assume extensive responsibility for the core processes. To do this, a needs analysis is first conducted to define which machines and systems are required, for example cutting machines, packaging machines, and centrifuges for separating fluids. *GEA* takes care of climate control at the factory so that food processing can take place under optimal conditions.

The customer relationship in this business model is characterized by the intensity and the longevity of the partnership *(**F**ar-reaching relationship)*. In each process step, large *GEA* project teams from the sales, technology, and engineering departments work intensively together with the customer as co-creators.

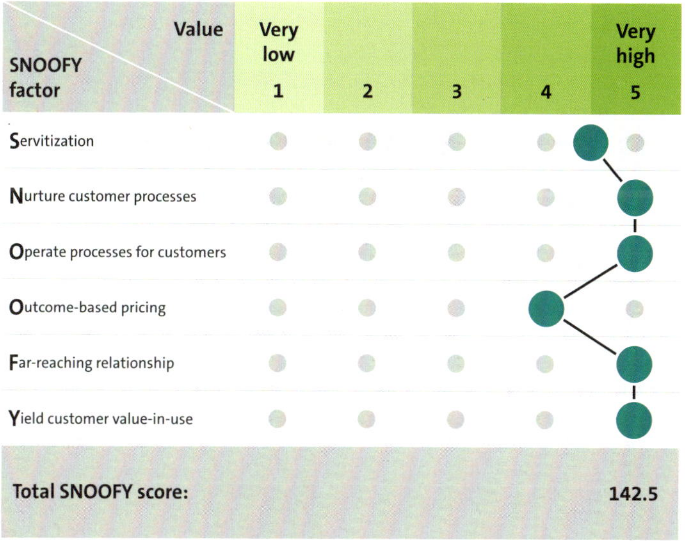

Business model profile no. 15: SAP – *Intelligent Enterprise* solution

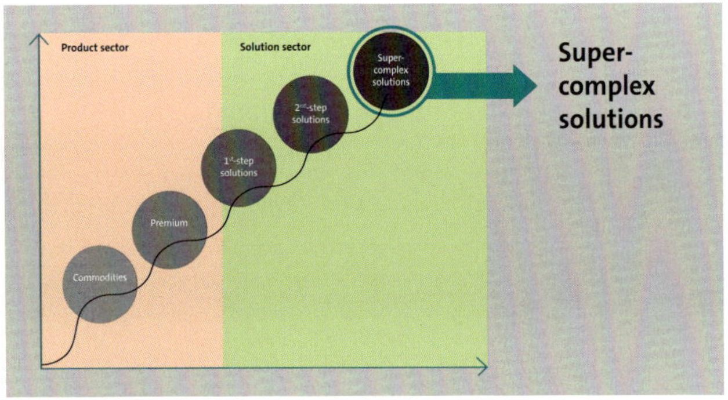

Company: SAP

Supplier of: Corporate software

Sales: $31 billion

Industry: Technology & software

What is marketed?	Intelligent enterprise solution for the digital transformation of all of a company's business processes
To whom is it marketed?	B2B customers from all industries
For which added benefit are customers willing to pay?	*"One of our most important distinguishing features is that we completely cover all of our customers' corporate processes using standard ERP software tailored specifically to the customer in the sense of Intelligent Enterprise. We are continuously evolving this due to the high innovation potential in the market. We aim to achieve long-term customer relationships so that we can set the course for the sustainable and successful digitalization of their business processes together with our customers."* Thomas Weinhardt – Vice President & Regional Sales Lead Purchasing Solutions for Medium-Sized Companies, SAP Germany

From tranquil Walldorf to 400,000 customers worldwide, *SAP* is Europe's biggest listed software company and the third-largest in the world. More than 100,000 employees and licensed partners work on innovative solutions and services each and every day. Since the company was founded in 1972, *SAP* has been developing standard software solutions for processing all business processes at companies (called ERP or enterprise resource planning solutions). These are rolled out at the customer's site by *SAP* or its partners in the framework of implementation projects.

In its most complex business model, *SAP* offers an *Intelligent Enterprise* solution that combines artificial intelligence and virtual reality. This solution is used to network all of the customers' business processes (e.g., CRM, ERP, SCM, SRM, HRM) end-to-end. It involves individually scalable software that *SAP* compiles specifically for every customer. As a result, the customer company is digitally transformed and the customer's business processes are standardized through the use of an all-encompassing software system. In this process, *SAP* digitally maps all of the customers' business processes. In addition, the customers' entire IT infrastructure is outsourced to *SAP* computer centers by means of cloud storage facilities.

The benefit to the customer arises due to the fact that all processes are mapped using the same digital system, thus increasing the efficiency of the work processes in the customer companies. By means of cloud services, *SAP* also assumes responsibility for its customers' entire IT and the related, highly sensitive data, as a result of which optimal functionality and efficiency are guaranteed *(**Y**ield customer value-in-use)*. The decision to invest in software, consulting, and operation is usually fostered through an optimal cost/benefit ratio that is determined in advance together with *SAP* and/or its partners as well as the customer *(**O**utcome-based pricing)*.

The integration of this complex solution into the customers' existing organization takes a long time. Together with the customer, *SAP* first defines which business processes are to be mapped using the solution. In the subsequent project phase, *SAP* checks how many workplaces are to be equipped with the software and which employees are to be able to access which processes, among other aspects. *SAP* then implements the system at the customer's site and sets it up *(**O**perate processes for customers)*. The complexity primarily involves integrating all of the previously used applications and processes, for example for CRM, into the new *SAP* software and linking them to one another.

Once the solution has been implemented on site, *SAP* or its partners remain the first point of contact for the customer. *SAP* carries out life cycle management for the solution independently. This includes service and maintenance on site as well as the option for the customer to contact *SAP* at any time *(**S**ervitization)*.

SAP works together with licensed partners in order to adequately look after the variety of customers and their individually adapted solutions around the globe. Both the partners and *SAP* itself attempt to obtain a deep understanding of the customers' value-creation processes in order to enable a bespoke solution. Intensive exchanges take place with the customers, especially during the initial phases. This intensive cooperation before, during, and after implementation leads to ***F**ar-reaching relationships* in this business model.

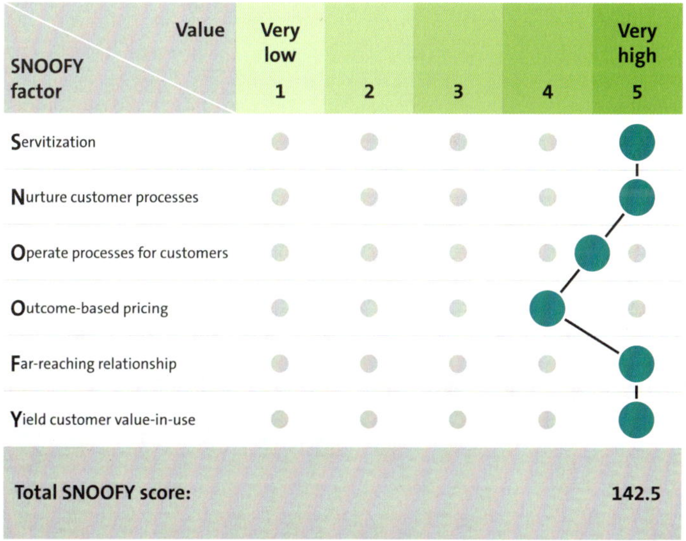

Business model profile no. 16: Schneider Electric – *EcoStruxure* solution for developing smart factories

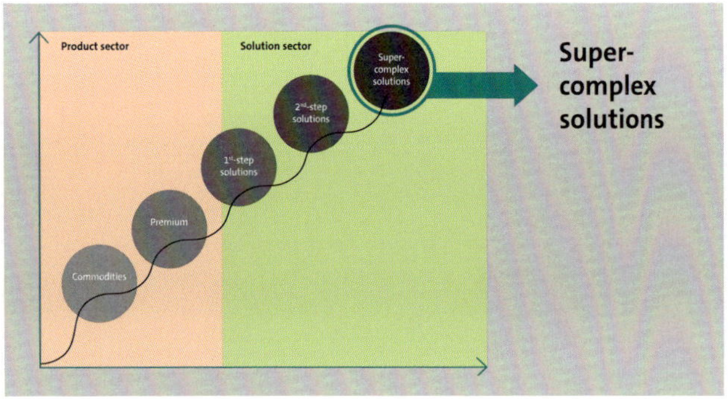

Company: Schneider Electric

Supplier of: Electrical engineering, energy distribution, industrial automation

Sales: $31.8 billion

Industry: Technology & software

What is marketed?	Development of smart factories
To whom is it marketed?	Industrial companies from various industries
For which added benefit are customers willing to pay?	*"With our EcoStruxure solution, we enable our customers to make the leap into Industry 4.0. It encompasses the digital transformation of their entire production value chain. As a result, their operating costs are reduced and their system operating rates are increased significantly."* Gerhard Wolf – Head of Sales Germany – Digital Solutions & IIoT at Schneider Electric

Schneider Electric is a globally operating electrical engineering group that is specialized in the business sectors of energy management and automation technology. Worldwide, more than 135,000 employees in more than 100 countries work on future-oriented industrial solutions.

The *EcoStruxure* solution business model is aimed at networking and automating customers' value chains, for example from production and packaging systems and machines to stockkeeping and logistics. Digitally mapping the corporate processes in the IIoT, which first requires networking the individual products and systems, is necessary to do this. In the *EcoStruxure* solution, *Schneider Electric* works closely together with its subsidiary *AVEVA*, which is specialized in industrial software.

EcoStruxure enables customer companies to structure their engineering and operational procedures more sustainably and profitably and to increase the performance of their systems, production, and supply chains. The solution enables production costs to be reduced by 15% and the system operating rate to be increased by 20% *(**Y**ield customer value-in-use)*.

An in-depth analysis of customer requirements marks the starting point of *EcoStruxure*. This is followed by consulting on the optimal specifications for the hardware (e.g., sensors and connectors) and software for the respective customer's value chain *(**N**urture customer processes)*. To this end, *Schneider Electric's* consulting teams work together with the customers on site. Among other things, they also provide advice on applications.

Schneider Electric then takes over the digital transformation of the core processes and installs all software and hardware components at the customer's site *(**O**perate processes for customers)*. *EcoStruxure* is subdivided into five areas: "Building," "Plant&Machine," "Grid," "IT," and "Power," each of which can be controlled as a digital cloud solution. *Schneider Electric* offers to operate the *EcoStruxure* components and analyze the efficiency of the machines and systems that are used, the personnel, and additional operational procedures. In the "Grid" area, for instance, *Schneider Electric* installs the control solution for digitally transforming the existing power grid and analyzes the procedures once it has been implemented. The company also networks all devices. The *EcoStruxure* platform that is used is an interoperable plug-and-play architecture and enables the automated recognition and configuration of the customers' devices, and thus the digital networking of the customer infrastructure, as part of the IIoT.

In its pricing, *Schneider Electric* takes several components into consideration, starting with the number of software licenses needed and the number of machines connected. Based on the specific customer data, *Schneider Electric* also prepares calculations concerning the increase in performance that can be anticipated due to the *EcoStruxure* solution, and it additionally takes these into account in the price *(**O**utcome-based pricing)*.

With *EcoStruxure*, the customer relationship extends over the complete system life cycle *(**F**ar-reaching relationship)*. While *EcoStruxure* is being set up, teams from *Schneider Electric* work intensively together with teams from the customer company. Contracts for long-term cooperations in after-sales services and further software developments are then concluded on the basis of service-level agreements.

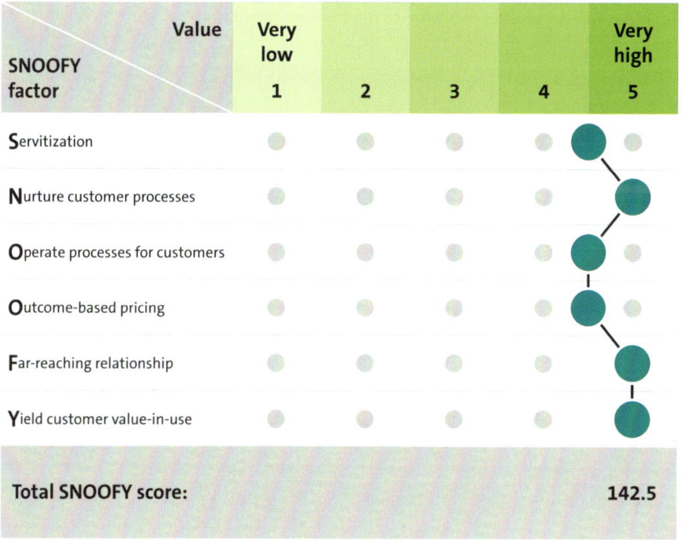

Business model profile no. 17: IBM – *Watson* solution

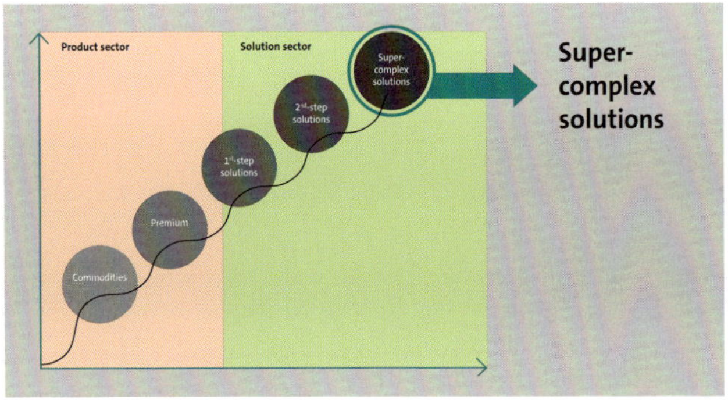

Company: IBM

Supplier of: Cloud and AI solutions as well as consulting services

Sales: $73.6 billion

Industry: Technology & software

What is marketed?	Increase in the efficiency of customer processes through the combination of AI, IoT, cloud services, and consulting
To whom is it marketed?	Large and mid-sized companies from various industries
For which added benefit are customers willing to pay?	*"The IBM Watson solutions can be integrated quickly and securely in many customer companies. We ensure that customer processes can run more efficiently and effectively using AI processes and automation. We combine all of this under the premise of maximum possible data and IT security."* Thorsten Schröer – Director Industry Business Development Industrial Market at IBM

IBM is one of the biggest cloud and AI groups with more than 350,000 employees worldwide. The company started out in the hardware business. Over the past 10 years, the core business has been extensively expanded to include multi-cloud, AI, and consulting and implementation services.

With its cloud solutions surrounding *Watson* and *RedHat*, *IBM* is now one of the market's leading providers of hybrid cloud services. The company operates the business model using a software-as-a-service approach. Customers can use it to portray and analyze data generated in the company clearly and in real time while only paying for computing power that is used. Artificial intelligence and cognitive machine learning additionally optimize the customer processes *(**Y**ield customer value-in-use)*. With the artificial intelligence of Watson, for instance, customers with their own manufacturing facility can reduce the effort that they would normally require for very laborious manual quality control by up to 80%. *Cognitive Visual Inspection*, for example, ensures sustainable product and process quality by analyzing products for faults using ultra-high-resolution cameras. *Watson* compares these images with the product images stored in the cloud or an edge computing solution (on-premise) and determines whether a fault is present or whether the product can be further processed. *IBM* guarantees the long-term functional capability of the solution. The pricing for this solution is based extensively on the use of the *Watson* solution by customers *(**O**utcome-based pricing)*.

With its *Watson* solution, *IBM* regards itself as the orchestrator of the various hardware and software components, services, and processes that are required *(**O**perate processes for customers)*. *IBM* provides the cloud platform and deploys the solution at the customer's site. *IBM* also operates its own local computer centers to administer the customer data securely and in conformity with data protection.

IBM additionally supports its customers with process analyses and comprehensive consulting *(**S**ervitization)*. To this end, the company has established the *IBM Global Business Service* business unit, which is represented in 170 countries and is the world's largest management consultancy. This business unit ensures that IoT solutions are adapted to individual customer requirements and according to customer feedback at all times *(**N**urture customer processes)*.

However, implementation does not yet mark the conclusion of the sales cycle. Together with its customers and partners, *IBM* works continuously on improving and operating the solutions. The intelligent systems are therefore extended and, if necessary, further AI processes (e.g., Acoustic Quality Inspection) are added. The exchange of communication is therefore very high during both the concept-creation phase and the long-term operating phase *(**F**ar-reaching relationship)*.

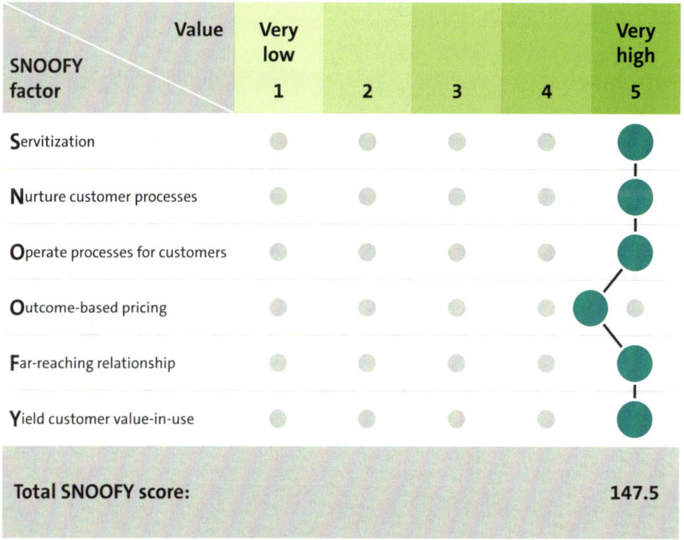

Business model profile no. 18: Siemens – Digital Enterprise Suite

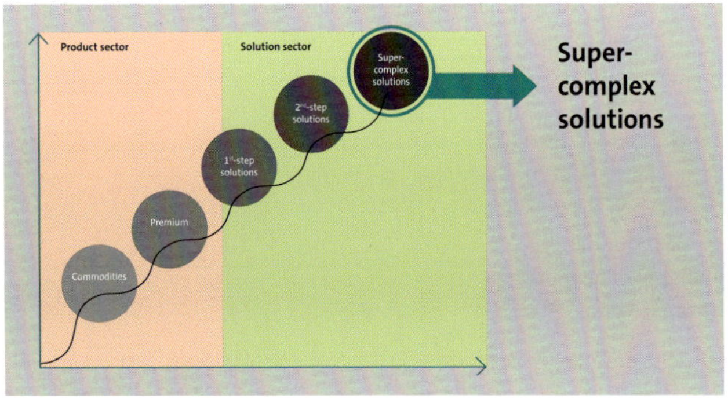

Company: Siemens

Supplier of: Various products, systems, and solutions

Sales: $66.8 billion

Industry: Conglomerate

What is marketed?	Digital transformation of the value chain
To whom is it marketed?	Companies and groups of all sizes and from all industries
For which added benefit are customers willing to pay?	*"We offer our customers an extensive portfolio of hardware, software, and services for industrial digitalization – with consistent tool chains from product design and simulation up to and including manufacturing planning and automation. Manufacturing data can be analyzed using an IoT platform in the cloud, and optimizations can be fed back into the processes. Simulations using the "digital twin" are an essential driver for speed: Product characteristics can be tested without ever having to build a sample, and manufacturing can be optimized long before a production facility has been commissioned in the first place. This allows users to automatically feed necessary manufacturing adaptations that affect the product design back into the process. All of that helps to gain valuable time before a new product is launched onto the market and to achieve the expected quality right at the start of production."* Hans-Jürgen Wamser – Vice President Sales Operations Digital Industries at Siemens

Siemens is a globally operating conglomerate with more than 290,000 employees in more than 290 countries. The group is a leader in the electronics and electrical engineering sectors and one of the world's highest-grossing companies in the industrial, energy, healthcare, as well as infrastructure and urban segments. In the course of the technological revolution, *Siemens* has specialized in the digital transformation of industrial companies.

With its *Digital Enterprise Suite*, *Siemens* offers Industry 4.0 solutions. In this, *Siemens* digitalizes its customers' value-creation process from product design to product manufacturing. To do this, *Siemens* uses a consistent concept of digital twins using the digital twins of the product, production, and the performance of the product and production. *Siemens* aims to establish a self-contained link between the virtual world of product development and production planning on the one hand and the physical world of the production system and product performance on the other hand. This link enables the findings obtained in the physical world to be used as aids for making informed decisions throughout the entire life cycle of the products and production processes. With the aid of simulations and prognoses concerning future performance, the digital twins enable an enormous added value ***(Yield customer value-in-use)***. For example, this process can be implemented to reduce the time to market of a new aircraft to as little as one-third of the usual duration. In the healthcare sector, the time from the development of vaccines to their availability on the market is dramatically reduced using the *Digital Enterprise Suite*. *Siemens* bases the pricing on its own expenditure and the benefit that arises for the customer when using the *Digital Enterprise Suite*. *Siemens* employs calculations based on use cases to demonstrate the possible impacts of the solution ***(Outcome-based pricing)***.

Siemens cooperates with its customers according to the "CIO principle": consulting, implementing, optimization. Here, *Siemens* takes over all processes that serve to integrate the entire product life cycle and subsequently enable continuous optimization ***(Operate processes for customers)***. All information and performance data available for customer production are uploaded to a cloud-based IoT platform *(Siemens MindSphere)* and made usable for the customer. *Siemens* additionally uses what is called *virtual commissioning* – digitally simulated commissioning of production – to analyze potential critical areas and errors in the customer's production.

Siemens also coordinates communication between the involved parties within the cloud and transfers the results of the simulations as far back as the customer's actual product design. *Siemens* also stands by its customers with consulting services and service hotlines in order to clarify any issues that arise as quickly as possible ***(Servitization)***. *Siemens* also offers workshops and training materials as part of this consulting.

Digital Enterprise Management establishes an all-encompassing digitalization and networking offer for a long-term customer relationship. In this case, the partnership between *Siemens* and its customers is a process of co-creation ***(Far-reaching relationship)***.

SNOOFY factor / Value	Very low 1	2	3	4	Very high 5
Servitization					●
Nurture customer processes					●
Operate processes for customers					●
Outcome-based pricing				●	
Far-reaching relationship					●
Yield customer value-in-use					●
Total SNOOFY score:					**147.5**

The fourth part of this book focuses on the **SALES** area of **SALESTEGY**. It is concerned with the effects of the business model steps on sales. It can be stated that entering a new business model step influences all areas of sales management.

Transferred to our imagery, the remaining chapters of this book can be classified as follows (see Figure 56).

Fig. 56 Chapter overview – Part 4

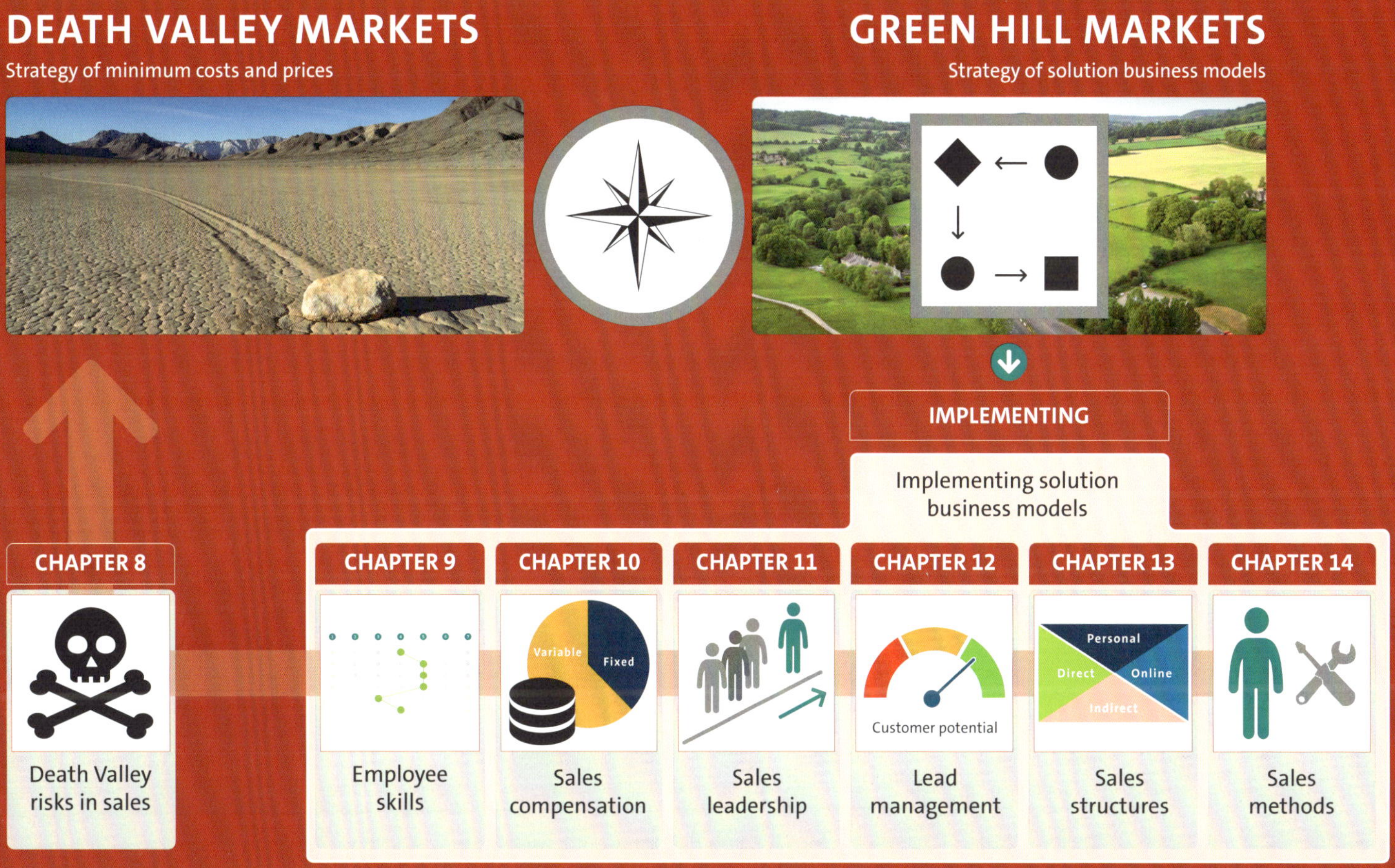

SALES – IMPLEMENTING NEW BUSINESS MODELS IN SALES

Chapter 8

Death Valley risks

Why new business models necessitate sales projects in six areas

Message of this chapter

Death Valley risks for new business models are lurking in sales.

What should you have learned by the end of the chapter?

- The Death Valley risks affect all areas of sales management, especially the following:
  - › Sales employee skills
  - › Sales compensation
  - › Sales leadership
  - › Lead management
  - › Sales structures
  - › Sales methods
- In order to counteract the Death Valley risks, I show sales management project steps for each of the six areas in Chapters 9–14.

Please note: When I talk about the implementation of *new* business models in this part of the book, I am referring to the implementation of new *solution* business models because, as described above, new business models in highly developed B2B markets are usually solution business models.

I introduce the individual implementation steps of sales projects in this part. This means that some points will be very detailed and that not every step is relevant to each reader. However, since the structure consists of the six project areas (sales employee skills, sales compensation, sales leadership, lead management, sales structures, and sales methods), you will be able to start at a point that is particularly relevant to you.

The following is not concerned with a defined sequence of projects in the six areas of sales. The numbering of Chapters 9–14 should therefore not be taken to mean that the employee skills always mark the starting point and sales compensation is then analyzed, etc.

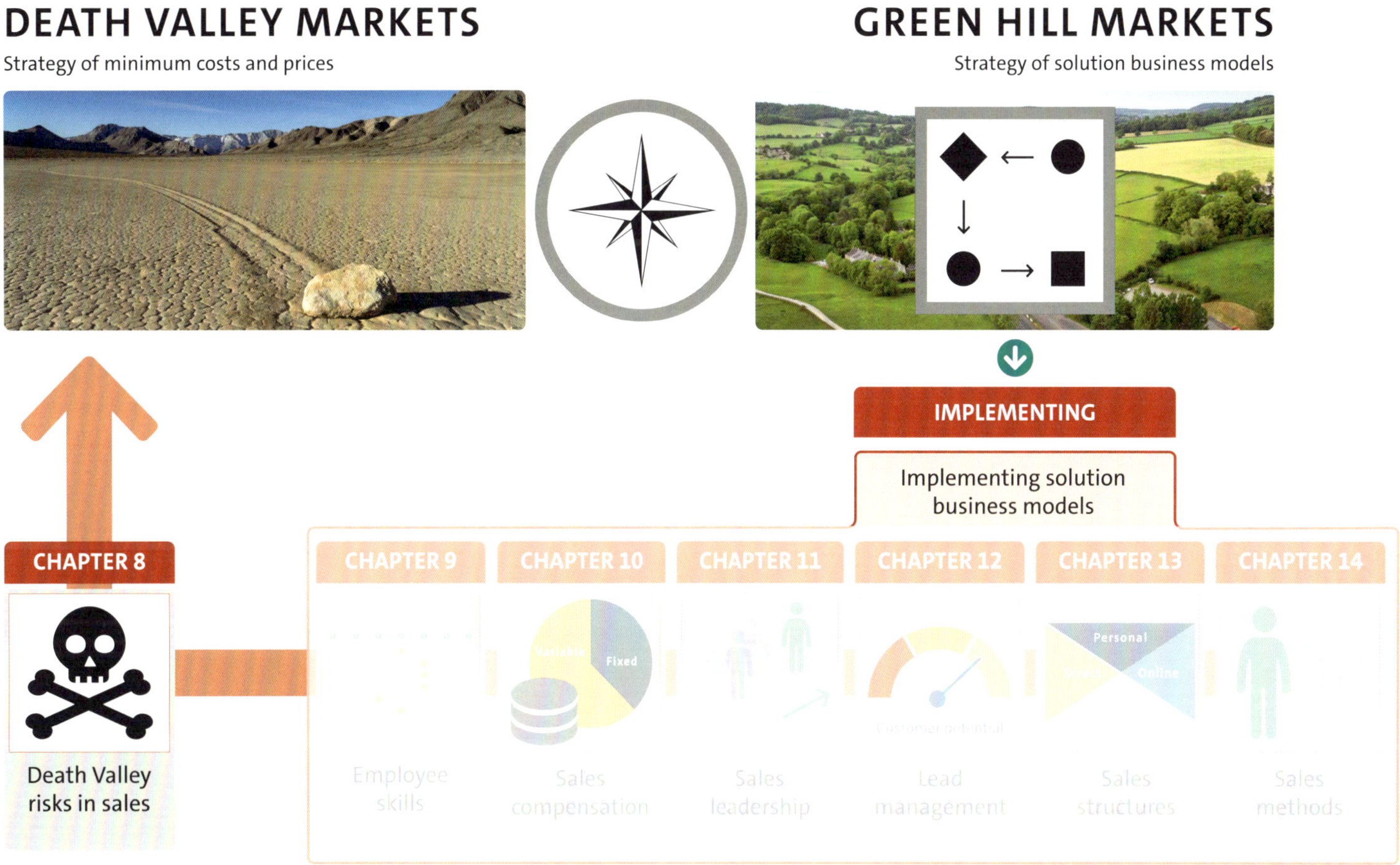
DEATH VALLEY MARKETS
Strategy of minimum costs and prices
GREEN HILL MARKETS
Strategy of solution business models
IMPLEMENTING
Implementing solution business models
CHAPTER 8
Death Valley risks in sales
CHAPTER 9
Employee skills
CHAPTER 10
Variable
Fixed
Sales compensation
CHAPTER 11
Sales leadership
CHAPTER 12
Customer potential
Lead management
CHAPTER 13
Personal
Online
Sales structures
CHAPTER 14
Sales methods

One central concept of this book is that many companies fail in sales when establishing solution business models. Figuratively speaking, Death Valley risks are lurking here because failing in sales casts you back into barren, low-margin market environments.

I have selected the six sales management areas that I go through in the following chapters because Death Valley risks are lurking there. The following list shows the Death Valley risks that we have observed most frequently in each case.

Death Valley risks in six areas of sales management

Sales employee skills

1st place: The skill profile for the sales employees was still originating from the product business.

2nd place: The necessary sales employee skill profile (target profile) for the new business model was not determined systematically.

3rd place: No training measures were undertaken to enable the sales employees to sell the new business model.

Sales compensation

1st place: The variable compensation was still oriented too extensively to the sales volumes in the classic product business.

2nd place: The variable compensation did not take the longer sales cycles in the solution business into consideration.

3rd place: The variable compensation did not offer any motivation for team selling.

Sales leadership

1st place: The leadership approaches were still originating from the product business.

2nd place: The ratio of first-line sales managers to sales employees was too high.

3rd place: The previously most successful product salespersons were selected as first-line sales managers in the new business model.

Lead management

1st place: The customer potential was overestimated.

2nd place: The customer potential for the business model was not calculated precisely enough (there was no differentiation according to customer segments, for instance).

3rd place: Lead management was not differentiated according to business model steps.

Sales structures

1st place: Establishing a new salesforce was shied away from.

2nd place: The weighting of the sales channels was not oriented adequately to the business model steps or the customer potential.

3rd place: Selling solution business models failed in individual countries and regions.

Sales methods

1st place: Classic sales methods were used too extensively, with the result that the necessity of the solution offer was not made clear to customers.

2nd place: No “Challenger Sale” method was applied.

3rd place: The sales method did not adequately take account of the risk that many customers perceive in the case of solution offers.

The Death Valley equation for failure in sales

I argue that each individual Death Valley risk is so powerful that it can cause an entire business model to fail. This hypothesis is expressed in the following **Death Valley equation** (see Figure 57). All six areas of sales management have to be set up correctly so that solution business models can prove successful (see equation, bottom line).

Fig. 57 The Death Valley equation for failure in sales

- If the right sales employee skills are lacking (1st line), new business models fail due to the lack of sales employee proficiency during implementation.
- If the right sales compensation is lacking (2nd line), new business models fail due to the sales employees' misdirected motivation.
- If the right sales leadership is lacking (3rd line), especially from first-line sales managers, there is both a lack of instruction for the sales employees in selling the new business model, which leads to a lack of sales employee proficiency, and a lack of adequate target-setting for the sales employees, which leads to a lack of motivation to sell new business models.
- If customer potentials are not determined systematically and the wrong target customers (leads) are approached as a result (4th line), the business model fails because the salesforce will be successful too infrequently.
- If an appropriate sales structure is lacking (5th line), customer access will prove to be the stumbling block. This is often the case when a more complex solution business model is to be sold using the same salesforce that has so far been responsible for classic product selling.
- If the appropriate sales method is lacking (6th line), customers will not recognize the benefit of even the most innovative solution. The stumbling block here often proves to be the unaccustomed needs determination along the process chains in the customer company.

We have experienced the risks in the six sales management areas time and again in our work together with our corporate partners. We have used these experiences to develop project steps to support companies in implementing their new solution business models in sales. Figure 58 provides an overview of this.

Fig. 58 Implementing new business models – project steps in sales management

	Chapter 9	Chapter 10	Chapter 11	Chapter 12	Chapter 13	Chapter 14
PROJECT	Sales employee skills	Sales compensation	Sales leadership	Lead management	Sales structures	Sales methods
STEPS	1. Defining relevant sales skills 2. Defining the target skill profile 3. Determining actual skills 4. Identifying gaps between actual and target profiles 5. Adapting personnel selection and development	1. Adapting compensation elements 2. Configuring variable compensation 3. Defining the payout curve	1. Setting sales targets 2. Guiding sales employees 3. Assessing sales employees	1. Determining customer data 2. Performing statistical analyses for determining customer potential 3. Integrating customer potentials in the CRM system 4. Implementing lead management per business model step	Decisions regarding 1. Sales channels 2. Establishment of a separate salesforce 3. Serving custo-mers in specific segments 4. Regional sales territories 5. International sales territories	Use of 1. SPIN selling methods 2. Challenger Sale methods 3. Customer inoculation in solution selling
RESULT	**Knowledge about:** Which sales employee skills do we need?	**Knowledge about:** Which sales compensation do we need?	**Knowledge about:** Which sales leadership do we need?	**Knowledge about:** Where is the greatest customer potential?	**Knowledge about:** Which sales structures do we need?	**Knowledge about:** Which sales methods do we need?

WHY SALESTEGY?

STRATEGY – FINDING THE RIGHT BUSINESS MODELS

PRACTICAL EXAMPLES

SALES – IMPLEMENTING NEW BUSINESS MODELS IN SALES

Chapter 9
Adapting sales employee skills

Project steps

- Step 1: Defining relevant sales skills per business model
- Step 2: Defining the target skill profile
- Step 3: Determining actual skills
- Step 4: Identifying gaps between actual and target profiles
- Step 5: Adapting personnel selection and development

Message of this chapter

The sales employee skills must be adapted for each new solution business model.

What should you have learned by the end of the chapter?

- You are familiar with the five project steps for adapting the skills of your sales employees.
- You are able to define the complex requirement profiles for each business model step.

DEATH VALLEY MARKETS

Strategy of minimum costs and prices

GREEN HILL MARKETS

Strategy of solution business models

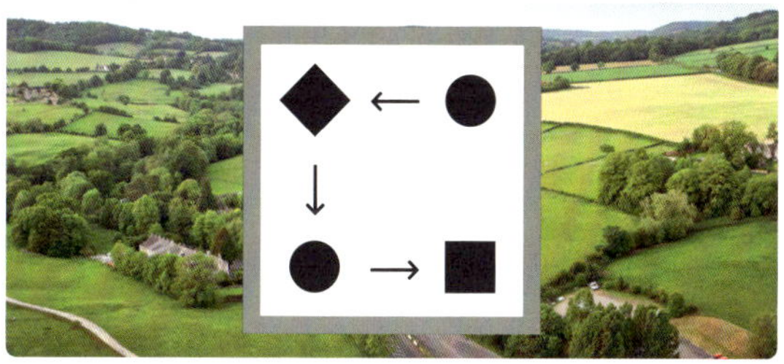

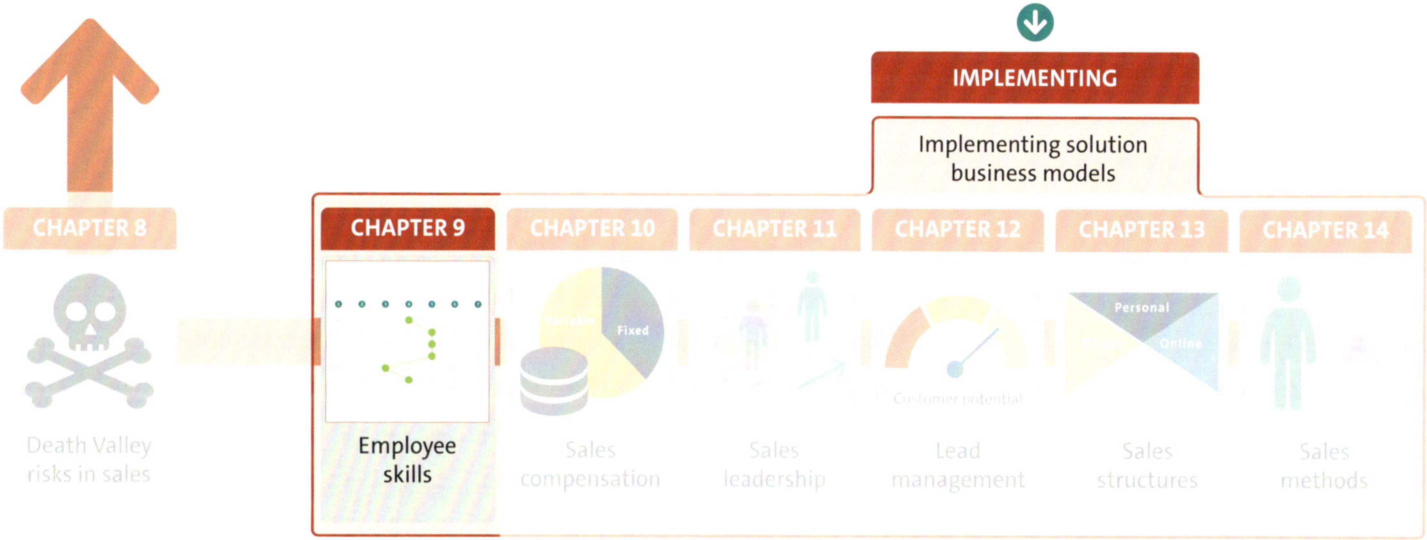

This chapter deals with the requirements that new business models make on the sales employees. Five steps are necessary to adapt the sales employee skills (see Figure 59).

Fig. 59 Implementing new business models – project steps in sales management

	Chapter 9	Chapter 10	Chapter 11	Chapter 12	Chapter 13	Chapter 14
PROJECT	Sales employee skills	Sales compensation	Sales leadership	Lead management	Sales structures	Sales methods
STEPS	1. Defining relevant sales skills 2. Defining the target skill profile 3. Determining actual skills 4. Identifying gaps between actual and target profiles 5. Adapting personnel selection and development	1. Adapting compensation elements 2. Configuring variable compensation 3. Defining the payout curve	1. Setting sales targets 2. Guiding sales employees 3. Assessing sales employees	1. Determining customer data 2. Performing statistical analyses for determining customer potential 3. Integrating customer potentials in the CRM system 4. Implementing lead management per business model step	Decisions regarding 1. Sales channels 2. Establishment of a separate salesforce 3. Serving custo-mers in specific segments 4. Regional sales territories 5. International sales territories	Use of 1. SPIN selling methods 2. Challenger Sale methods 3. Customer inoculation in solution selling
RESULT	**Knowledge about:** Which sales employee skills do we need?	Knowledge about: Which sales compensation do we need?	Knowledge about: Which sales leadership do we need?	Knowledge about: Where is the greatest customer potential?	Knowledge about: Which sales structures do we need?	Knowledge about: Which sales methods do we need?

Sales employees represent the central interface between the supplier and customer and are therefore among the most important multipliers of business model initiatives. One of the key tasks of sales employees in this is to get to the heart of the new business model and convince customers of its innovative value contributions. Business models that are inadequately implemented by the salesforce are doomed to fail. It is therefore elementary, particularly when introducing solution business models, for the sales employees to have or be able to develop the relevant sales skills.

Many sales employees who were successful in the original business fail when selling solutions. In an example using companies that my colleagues Werner Reinarz and Wolfgang Ulaga researched, only around one-third of sales employees successfully manage to progress from classic product selling to solution selling.[17]

Our own rough estimates arrive at a similar result (see Figure 60). According to them, 60% of the salesforce fail when switching from the product business to selling 1st- and 2nd-step solutions. As many as 75% fail when progressing from 1st- and 2nd-step solutions to super-complex solutions.

Fig. 60 The shift to solution business models – a high obstacle for the salesforce

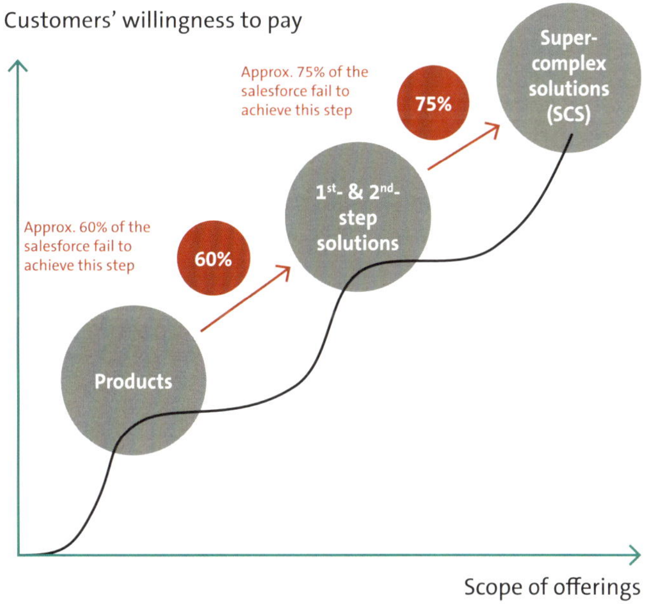

[17] Ulaga, Reinartz (2011), *Hybrid Offerings: How Manufacturing Firms Combine Goods and Services Successfully*, Journal of Marketing, 75 (11), p. 21.

SALES

We attempted to determine the reasons in a study. In this, we looked at the sales skill requirements in 510 companies from 12 industries.[18] Figure 61 shows the five most frequent deficits of sales employees. Firstly, it is interesting to note that there is obviously hardly any lack of technical/functional knowledge. Such deficits were only found in 1.4% of cases. Instead, the most frequent deficits are found in acting strategically, the ability to solve problems creatively as well as customer analysis and customer understanding. These skill deficits provide initial indications for explaining the high failure rate on the road to the solution business.

Fig. 61 Sales employee skill deficits

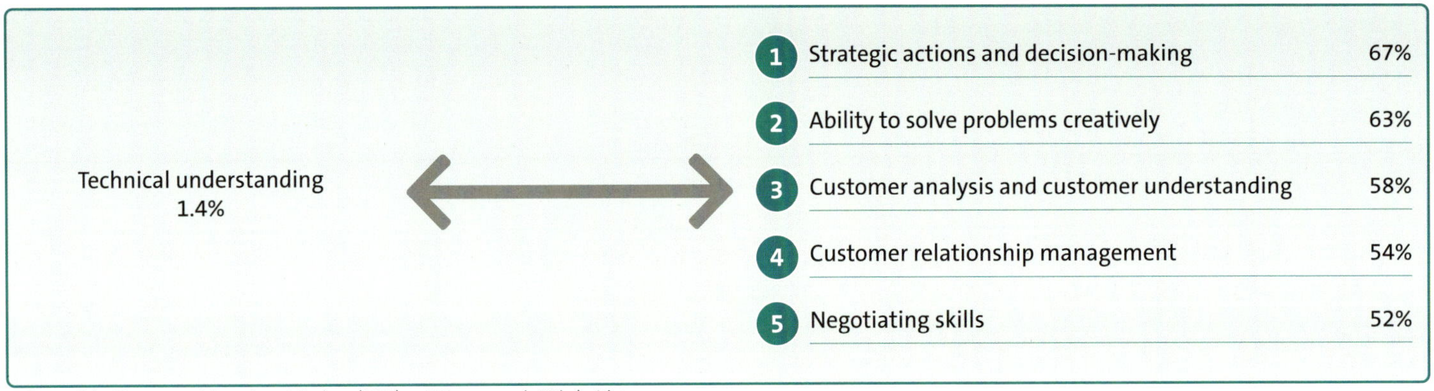

Source: Sales Management Department, Bochum (2015); n=510 managers in 12 industries

[18] Wieseke, Schmitz (2015), *Focus Study Concerning Skill Requirements in Sales*, Sales Management Department, Ruhr University of Bochum.

In order to meet the skill requirements on the road to the complex solution business, we have developed five project steps for skill assessment and development in cooperation with human resources (HR) managers. The five steps are shown in Figure 62. I deal with them in the following.

Fig. 62 Five steps for evaluating and developing sales employee skills

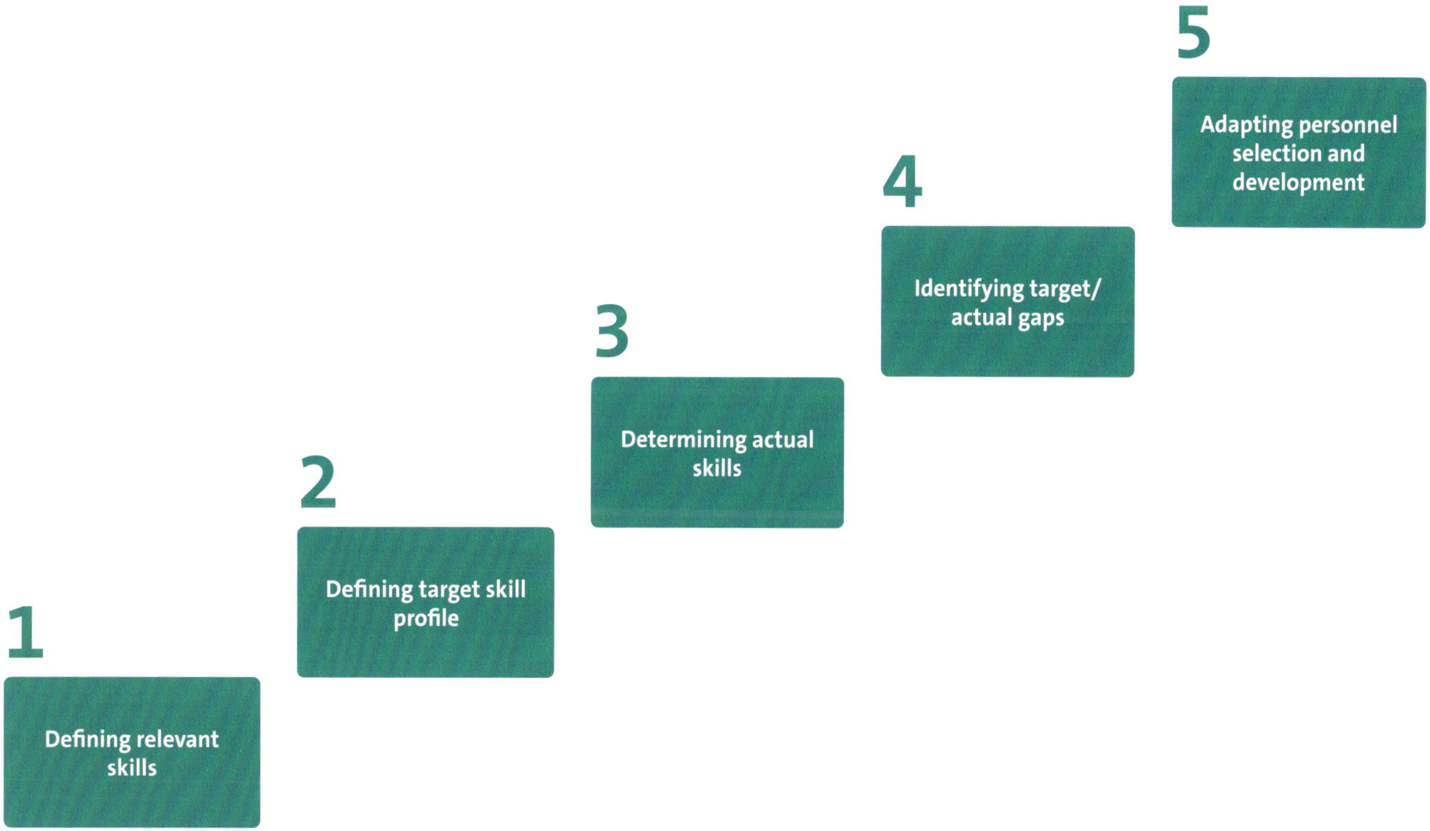

Project steps

Step 1: Defining relevant sales skills per business model

In order to successfully implement solution business models, it is first vital to determine precisely which sales employee skills are important in the respective business model. To accomplish this, we have developed skill requirement profiles for each business model step.

A precise overview of the skills is shown in Figure 63. The figure shows that the individual skill requirements can be subdivided into four areas:

1. Technical skills
2. Methodological skills
3. Communication skills
4. Social skills

I describe the differences in the skill requirements for the five business model steps in the following. It is important for me to point out that the figure shows **average** requirements. However, each business model has its own requirements. So although the values in the figure are helpful as an initial starting point, they are no substitute for a precise analysis for each individual business model.

Fig. 63 Skill requirements for sales employees per business model step

		Commodities	Premium	1st-step solutions	2nd-step solutions	Super-complex solutions	Key accounts
Role of the sales employees		Order taker	Premium supplier	Service explainer	Solution consultant	Strategic consultant	Key account manager
Undertaken by		Individual	Individual	Individual/team	Team	Team	Individual/team
Requirement of sales employee		Product knowledge	Knowledge about the use of the product by customers	Knowledge of the customer's business model and the value that the products/ services create for it	Knowledge of the respective industry and value chains and how solution offers create a value contribution for the customer	Strategic analysis of the neuralgic points of the customer business model and joint solution development	In-depth knowledge of the customer organization. Knowledge of how customer satisfac-tion and loyalty can be ensured in the long term
1. Technical skills	Product & technical knowledge						
	Recognizing customer needs						
	Developing solutions for the customers' value chains						
	Knowledge of the industry				*More in-depth knowledge of the industry and the business models*	*More in-depth knowledge of the industry and the business models*	

		Commodities	Premium	1st-step solutions	2nd-step solutions	Super-complex solutions	Key accounts
Role of the sales employees		Order taker	Premium supplier	Service explainer	Solution consultant	Strategic consultant	Key account manager
Undertaken by		Individual	Individual	Individual/team	Team	Team	Individual/team
1. Technical skills	Academic background	○	○	◔ *Calculating the business case*	◑ *Skill combination in the team consisting of:* • *Engineering know-how* • *Commercial knowledge*	◕ *Skill combination in the team consisting of:* • *Engineering know-how* • *Commercial knowledge*	◔
	Digitalization expertise	○	○	◔	◑	◕	◔
2. Methodological skills	Administration	◔	◔	◔	◔	◔	◔
	Self-organization	◔	◔	◑	◕ *Self-organization in the team*	◕ *Self-organization in the team*	◕
	Strategy development	○	○	◔	◑	◕ *Long-term strategy development together with customers*	◕
	Pricing skills	○	○	◔	◑	◕	◕
	Target-setting & evaluation of own performance	◔	◔	◑	◕	◕	◕

		Commodities	Premium	1st-step solutions	2nd-step solutions	Super-complex solutions	Key accounts
Role of the sales employees		Order taker	Premium supplier	Service explainer	Solution consultant	Strategic consultant	Key account manager
Undertaken by		Individual	Individual	Individual/team	Team	Team	Individual/team
3. Communication skills	Question techniques & conducting sales calls		*SPIN selling (see Chapter 14)*	*Challenger Sale (see Chapter 14)*	*Challenger Sale and consultative selling*	*Challenger Sale and consultative selling*	
	Negotiation skills						
4. Social skills	Empathy						
	Team skills						
	Networking skills						
	C-level standing						

Commodity business models

The fundamental requirement in commodity business models is to get the available product to as many customers as possible. Since these products are usually easy to handle, there is hardly any need for explanation. Sales is focused on knowledge of the product and the effort of the sales employees. We describe their role as **order takers**. It encompasses the following requirements:

1. **Technical skills:** Order takers must have the necessary product knowledge to be able to explain the details of the product to customers.
2. **Methodological skills:** It is important for commodity business model sales employees to be able to carry out both the fundamental administrative activities and the tasks surrounding the purchase process professionally. They must be able to write an offer, organize their journeys to the customers, and use digital sales tools (handling of CRM systems, for example) to be able to process a lot of customers as quickly as possible.
3. **Communication skills:** The order taker needs a minimum level of communication and negotiation skills.
4. **Social skills:** Empathy on the part of the sales employees is the prerequisite for being able to recognize and interpret customer needs during sales calls. As can be seen, team skills, networking skills, and C-level standing play a comparatively subordinate role.

Premium business models

In the premium business model sector, the high quality and the related benefit of the product form the focus of the interaction between sales employees and customers. Customer loyalty also increases in importance. Here, the sales employees take on what is called a **premium supplier** role. It encompasses the following requirements:

1. **Technical skills:** It is important for the suppliers of premium products for sales employees to have both a detailed knowledge of the product and an overview of the requirements of their customers. Both are necessary to establish high customer loyalty in the long term.
2. **Methodological skills:** Besides the fundamental methodological skills that are already important in the commodity business models, the premium supplier is focused on ensuring smooth processes. This fosters longer-term customer loyalty and cultivates an image of a high-quality supplier.
3. **Communication skills:** The requirements of question techniques, dialog skills, and negotiation skills increase in comparison with commodity selling. Sales employees involved in premium sales must have the ability to use benefit-centered sales methods such as SPIN selling (see Chapter 14).
4. **Social skills:** In order to justify the higher price of premium products through an adequate customer benefit, the sales employees must be able to put themselves in the customers' position and recognize their needs as early on as possible. A high level of empathy is therefore the basis for a successful customer relationship. The networking skills of the sales employee are also useful for building up an expert network within which they can place their premium products.

1st-step solutions

1st-step solutions are no longer concerned solely with the product, but with developing more extensive solutions for the customer. To do this, the sales employees must be familiar with their customers' processes. They must be able to register the customers' needs and independently show them the options that their solutions can provide. The role of the sales employees can be described as that of a **service explainer** here. It encompasses the following requirements:

1. **Technical skills:** Product knowledge remains an important factor. Sales employees must also know the services to which the product offers can be linked and how these solutions can ultimately provide the customer with added value. To do this, they must be familiar both with the customers' situation and with the significant changes and trends in the respective industry. The sales employees additionally require commercial background knowledge. It enables them to calculate business cases together with the customer and therefore to demonstrate the added value of the solutions on the basis of figures.

2. **Methodological skills:** One major difference compared to the product business models is in self-organization as well as the ability to set targets and evaluate one's own performance. This is due to the complex and protracted planning of the sales processes. In the case of 1st-step solutions, considerably more customer visits with diverse contact persons usually have to be organized in order to conclude a sale.

3. **Communication skills:** Communication skills are a crucial element of successful selling on this business model step. The customers' relevant requirements have to be identified even more extensively than before using question techniques. According to the Challenger Sale concept (see Chapter 14), the customers' previous processes additionally have to be "challenged." The sales employees also have to anticipate future problems here and discuss them with the customer. A good feel for how to conduct the sales calls with the customers is required to do this. During the introductory phase of 1st-step solutions, these capabilities often still need to be established in large parts of the salesforce.

4. **Social skills:** Along with conducting sales calls empathetically, team skills, networking skills, and C-level standing take on increasing importance. Consequently, the sales employees must be able to establish a corresponding network within both the customer's and their own organization. C-level standing means that the sales employees can communicate on an equal footing with the managing directors of the customer companies, since this is usually where the final decision on purchasing 1st-step solutions is made.

2nd-step solutions

Intervention into the customer value chain becomes even more extensive in 2nd-step solutions. Together with the customers, the sales employees develop long-term process improvements that provide the customer with a permanent value contribution in this case instead of delivering only temporary solutions. The role of the sales employees can be described as that of a **solution consultant** here. It encompasses the following requirements:

1. **Technical skills:** Product knowledge still plays a role in 2nd-step solutions, but its importance decreases compared to the previous business model steps. It is now more relevant for the sales employees to generate knowledge about the customer processes in order to be able to develop solutions together with the customer. They must also know their way around the respective industry so that they can offer the customer an added value as "knowledge brokers." Work is increasingly carried out in sales teams. The skills have to be distributed in a well-balanced way within the teams so that both engineering knowledge and commercial expertise are covered.
2. **Methodological skills:** One major difference compared to the previous business model steps is self-organization within the team. It increases in importance because sales employees with different capabilities work together. The demands on joint target and strategy development also increase. Sales employees must be able to develop a market processing strategy and derive set targets from it within the team.
3. **Communication skills:** As in 1st-step solution business models, sales employees must also possess Challenger Sale capabilities (see Chapter 14) in this case. They are also no longer merely sellers of solutions, but increasingly bring their consulting skills into play in the customer company in the sense of "consultative selling."
4. **Social skills:** Besides empathy and team skills, the importance of the sales employees' networking skills and C-level standing also increases. They must be able to discuss complex issues with the customer and make themselves heard on higher hierarchy levels. An extensive network of experts is crucial for the quality of the consulting service that the sales employees provide in 2nd-step solutions.

Super-complex solutions:

The pinnacle of complexity and customer relationship longevity is reached in super-complex solutions. Sales employees become **strategic consultants** who have to integrate the central business model processes of suppliers and customers. Both a C-level standing and the ability to scrutinize the customer processes are critical here to be able to convey the added values to the customer. Sales employees act as strategic consultants and often also as digitalization experts who convey competition-critical solutions to the customers using their knowledge and experience. The role of **strategic consultant** encompasses the following requirements:

1. **Technical skills:** The technical skill profile for sales employees in super-complex solutions is similar to that for 2nd-step solutions. They also act as knowledge brokers here and develop complex solutions together with the customer. The profile is rounded off by expertise in the area of digitalization.

2. **Methodological skills:** Again, organizational skills within the sales team are crucial to success. In addition, the employees now have to be able to structure the long-term strategy more intensively together with the customer. The ability to calculate prices for super-complex solutions is another important skill dimension.

3. **Communication skills:** The Challenger Sale question techniques (see Chapter 14) are also helpful here to highlight the value of solutions. The communication skills of the strategic consultants are comparable to those of 2nd-step solution sales employees.

4. **Social skills:** A high-quality expert network is even more important for super-complex solutions than in the previous business model steps. Different specialists, who are involved on a project basis, are required to accompany the customer's process development accordingly. The entire sales team additionally requires a high degree of stamina for long-term strategy development with the customers in order to remain present during the long sales cycles. Empathy and team skills as well as a high level of C-level standing are crucial for this.

Information box: Special case of key account managers

The skill profile of key account managers is a special case (see Figure 63, right-hand column). They are responsible for significant sales volumes and must therefore satisfy a demanding skill profile. On the whole, this profile is similar to that of the strategic consultants, but there are some deviations:

1. **Technical skills:** Just like strategic consultants, key account managers must also possess good industry, product, and technical knowledge in addition to often needing academic background knowledge. The skill to determine needs must also be extensively pronounced. Since many key account managers are primarily responsible for selling simple products and solutions and not for 2nd-step solutions or super-complex solutions, their solution skill requirements are lower than those of strategic consultants on the whole. The same also applies to digital expertise.

2. **Methodological skills:** In order to structure the processes as smoothly as possible for the customer, key account managers need extensive administration and organizational skills. They also require strategy development, pricing, independent goal-setting, and evaluation abilities. They can only establish a successful and close relationship with the customer in the long term if they have these skills. However, the processes are not as complex as in super-complex solution selling.

3. **Communication skills:** In order to recognize customers' needs and maintain the relationship, question techniques, negotiation skills, and good conversational skills are equally as relevant for key account managers as for strategic consultants.

4. **Social skills:** In terms of team skills, the average requirement is lower than that for strategic consultants, as not all key account managers work in large teams. Customer relationship management is one of their main tasks; focus therefore shifts to empathy and effective communication with the customer. The requirement profile here is on a par with that of the strategic consultants.

Step 2: Defining the target skill profile

Once the core skills have been defined, the HR department can create target profiles. They show the skill profile that is relevant for the respective position. The requirements of the respective business model should be incorporated into these. The target profile must therefore be developed to reflect the requirements.

Figure 64 shows an example of a target profile for a 1st-step solution business model for a telecommunications provider. This excerpt refers to the technical skills.

Fig. 64 Technical skill requirements for sales employees (example)

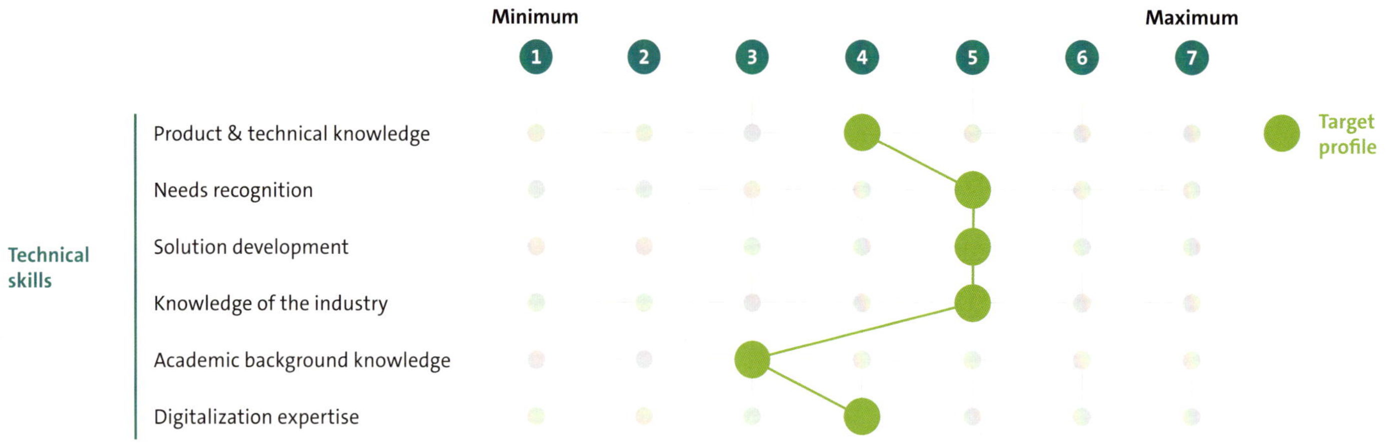

Example: 1st-step solution business model of a telecommunications supplier

Step 3: Determining actual skills

In the third step, the actual status of the relevant skills in the existing salesforce is determined. Depending on which target skills have been defined, corresponding assessment tools must be used to record the employees' skills. HR departments can use classic assessment centers for this. They can also use 360° surveys in which self- and external perceptions of employees, supervisors, and customers are interlinked. The HR department can also use standardized test methods for specific skills such as analytical capabilities.

Step 4: Identifying gaps between the actual and the target profiles

In the fourth step, the HR department has to determine the gaps between the actual and the target skill profile of the existing salesforce. Figure 65 continues the above example of the telecommunications provider and compares the actual and the target profiles for the technical skills. As can be seen, the biggest challenges in the relevant salesforce lay in the areas of needs recognition, solution development, and industry knowledge.

Fig. 65 Nominal/actual comparison of technical skill requirements (example)

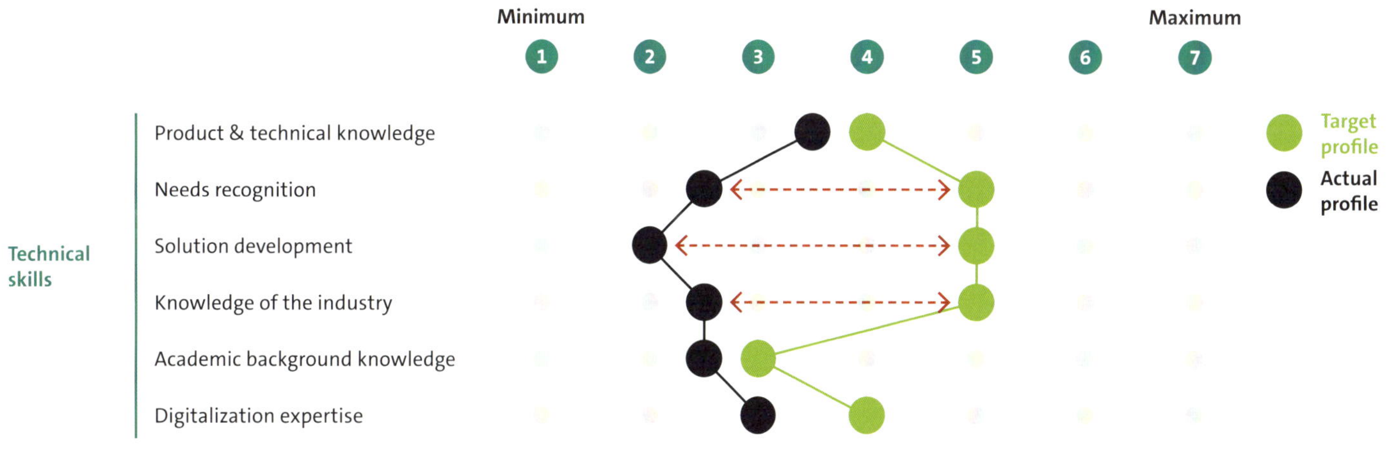

Example: 1st-step solution business model of a telecommunications supplier

Step 5: Adapting personnel selection and development

The fifth step deals with the development areas required to achieve the target skill profile. Two options are possible for filling the gaps between the actual and the target profiles: Personnel development or personnel selection.

The classic personnel development tools can be used for sales employees within the existing salesforce who are sufficiently capable of development. This particularly refers to training and coaching programs. Examples of these include programs such as the *Miller-Heiman Method*, *Six Battlefields* from *MERCURI International*, and *the Clean Sales Method* from *Buhr & Team*.

As described above, the number of sales employees who are able or willing to make the switch to solution selling is often insufficient. In this case, suitable employees then inevitably have to be recruited externally. When selecting personnel, it is then important to develop methods that test the skills corresponding to the target profile. Demanding selection procedures are necessary due to the broad range of skills required. In particular, "multimethod" assessment procedures have proved suitable for this. They are able to test a variety of skills.

Both training and personnel selection processes should be evaluated after completion. This evaluation can determine whether the steps taken have achieved the desired effect or whether readjustment is necessary. The actual situation must then be reviewed again to ascertain whether the training measures and the recruitment of new sales employees have proved worthwhile. The previously identified gaps should close. It is also important to continually review the previously identified core skills because the skill requirements also change continuously when the business models are adapted.

WHY SALESTEGY?

STRATEGY – FINDING THE RIGHT BUSINESS MODELS

PRACTICAL EXAMPLES

SALES – IMPLEMENTING NEW BUSINESS MODELS IN SALES

Chapter 10
Adapting sales compensation

Project steps

- Step 1: Adapting compensation elements
- Step 2: Configuring variable compensation
- Step 3: Defining the payout curve

Message of this chapter

The sales compensation has to be adapted for each new solution business model.

What should you have learned by the end of the chapter?

- You are familiar with the three project steps for adapting the sales compensation for your solution business models.
- You are able to assess the complex requirements of the variable compensation.

DEATH VALLEY MARKETS

Strategy of minimum costs and prices

GREEN HILL MARKETS

Strategy of solution business models

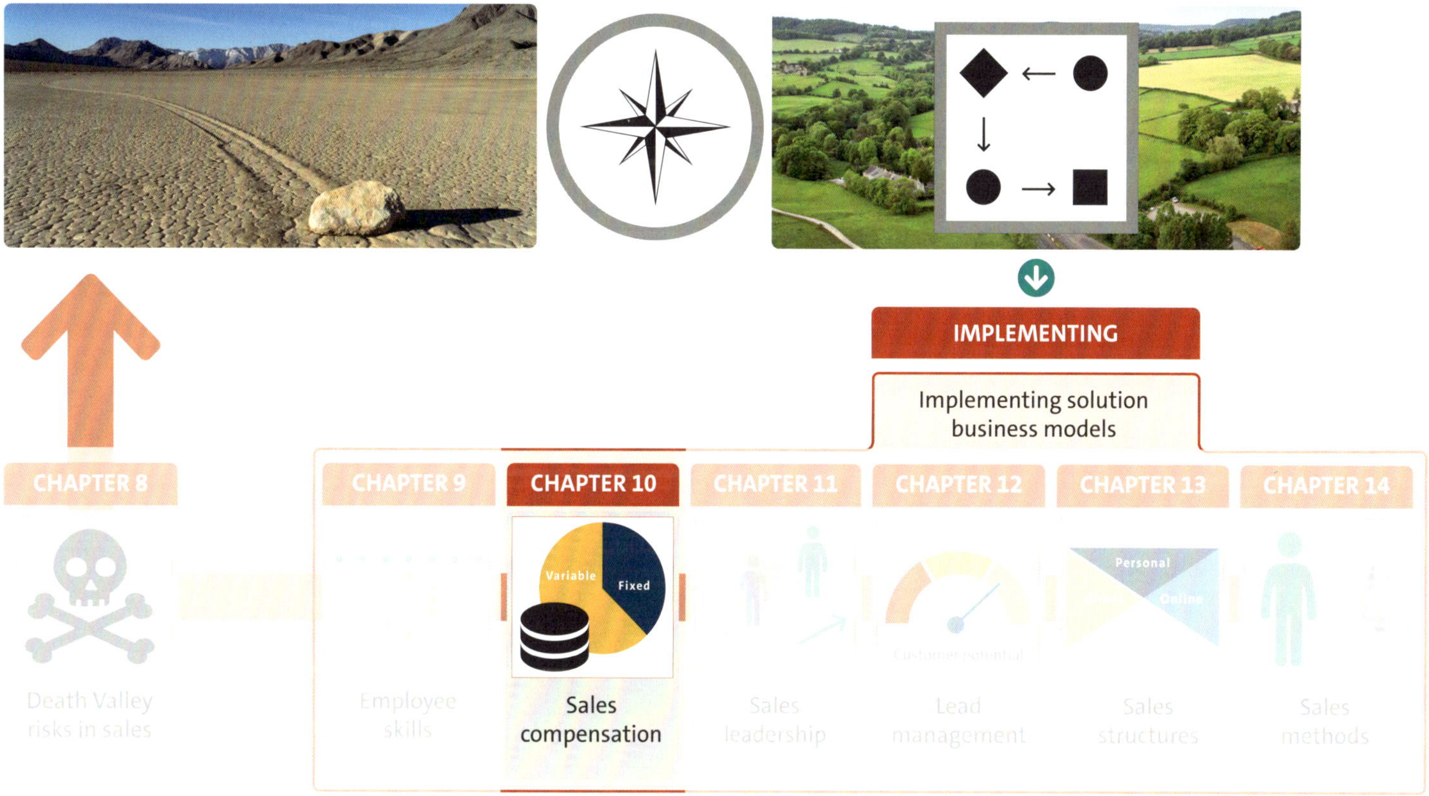

In this chapter, I deal with the sales compensation. Three project steps are necessary to adapt it as well as possible to the requirements of new business models (see Figure 66).

Fig. 66 Implementing new business models – project steps in sales management

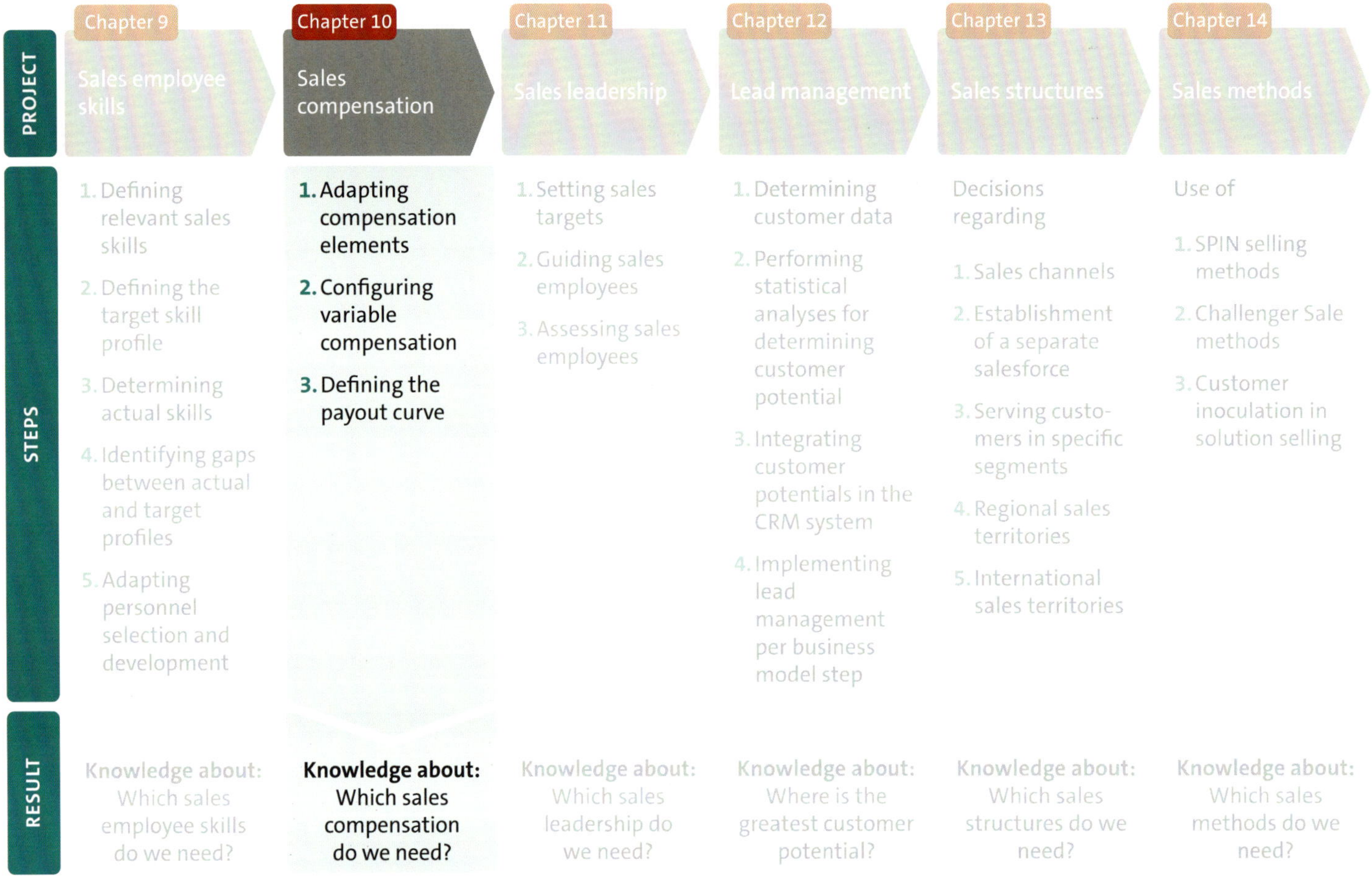

Many solution business models fail due to the fact that the motivation to sell solution offers suffers due to unadapted compensation. Opinions especially differ as regards the variable compensation. This was already the case in classic product selling, where advocates of variable compensation swear by its motivational effect, but opponents warn of the erosion of intrinsic motivation. The distance separating the two camps as regards the evaluation of variable compensation in solution selling is at least equally as wide.

Figure 67 shows a dialog that makes this conflict tangible. I experienced it at a sales compensation roundtable. The two opponents are sales directors at different mechanical engineering companies, both of which are on the road to becoming solution suppliers. As can be seen, the advocate for variable compensation is of the opinion that sales cannot manage without variable compensation, whereas the risks related to variable compensation in the new solution business models are dominant for the skeptic.

Fig. 67 Two perspectives regarding variable compensation in the solution business

+

Advocates for variable compensation:

"Variable compensation remains important to motivate sales employees."

"That's why the variable compensation has to be linked to the sales employees' sales volume in the solution business."

"But no, variable compensation isn't a solution either. Without variable compensation, motivation is lower. Ultimately, it's all about the team moving forward, even if it occasionally overshoots the target here and there."

−

Skeptics of variable compensation:

"Yes, but variable compensation can quickly steer sales employees in the wrong direction in the solution business. Traditional compensation approaches no longer function here."

"That alone doesn't help, though, because if the variable compensation component of the total compensation is too high, the sales employees constantly work flat-out. That might be good for the traditional product business, but not for developing complex solutions with the customers. That takes time."

"I'm more critical. Solution selling is far more about working smart than about working hard and fast. If the variable compensation fosters working hard and fast too much, it comes at the expense of working smart, and sometimes slower."

These two standpoints are also reflected in the compensation models implemented in sales. Whereas some companies have extensively shifted their compensation models for the solution business in the direction of a high basic salary, others still swear by high variable components.

So what is to be done if your company is on its way toward the solution business? As you may have already assumed, there is no general answer to this question. The respective parameters of the business model, the company, and the market have to be taken into consideration. The discussion therefore has to be broken down to the individual controlling levers first of all. Only then can clear recommendations be issued.

A total of nine controlling levers are relevant for adapting the compensation model. They are shown in Figure 68. We will take a closer look at them in the following. In compensation projects, the nine controlling levers can be assigned to three steps:

1. Adapting compensation elements
2. Configuring variable compensation
3. Defining the payout curve

Are nine controlling levers really necessary for the right compensation? It is important to me to point out the following: These nine controlling levers are not just theory. Each year, we work together with companies on their sales compensation at *Prof. Schmitz & Wieseke – Sales Management Consulting*. In this work, we discover time and again that companies get into difficulties if they disregard one of the nine controlling levers. So, I argue that a compensation model that has not been thought out properly will only look good on paper, but it will fail in practice. It will fail either because it is too expensive for the company or because it does not motivate the salesforce adequately.

Fig. 68 The nine controlling levers of variable compensation depending on the business model step

Controlling lever	Explanation	Business model Commodities	Premium	1st-step solutions	2nd-step solutions	Super-complex solutions
Step 1: Adapting compensation elements						
1. Salary level	Level of total compensation	Salary level lower than in solutions				Salary level higher than in products
2. Variable compensation component	Share of compensation influenced by performance	Variable share higher than in solutions				Variable share lower than in products
3. Further compensation elements	E.g., handling of new recruits; handling of days absent; sales competitions (sales contests)	E.g., individual competition for the most sales	E.g., individual competition for the most sales	Various types of competitions	E.g., team competition for the best project	E.g., team competition for the best project
Step 2: Configuring variable compensation						
4. Performance measurement	Defining the assessment basis for the variable compensation	• Sales volumes and gross margins with commodities • Short-term targets • Individual performance	• Sales volumes and gross margins with premium products • Short-term targets • Individual performance	Individual and team components, e.g. number of service contracts	• Sales volumes and gross margins with solutions • Long-term targets • Team performance	• Sales volumes and gross margins with solutions • Long-term targets • Team performance
5. Choice between bonus and commission	Bonus: Fixed amount on achieving targets; commission: percentage payout per product sold	Both possible	Both possible	Bonus more likely than commission	Bonus more likely than commission	Bonus, no commission
6. Payment cycle	Frequency of compensation payouts	As often as possible	Frequent, small payouts	More rarely, large payouts	Rare, large payouts	Rare, large payouts
Step 3: Defining the payout curve						
7. Start of variable compensation	Target achievement level at which the variable compensation begins	Early start of variable compensation				Late start of variable compensation
8. Start of over-performance factor	Is there to be an over-performance factor and, if yes, at which target achievement level?	Early start of over-performance factor				Late start of over-performance factor
9. Form of over-performance factor	Change in payout curve gradient after target achievement	Sharp increase		Slight increase		No increase

Project steps

Step 1: Adapting compensation elements

Let us first examine the compensation elements. This is concerned with the salary level, the variable compensation component, and further compensation elements.

1. Salary level

The salary level describes the overall level of the salary, consisting of all fixed and variable components. The salary level in the solution sector should be higher on average than in the product sector. The background to this is that the skill requirements discussed in the above for the sales employees increase as the level of complexity grows and that the employees have to be compensated accordingly.

2. Share of variable compensation

The variable component of the salary is paid depending on the performance of the respective sales employee. As described in the sales leadership section (see Chapter 11), the sales employees in the product business tend to be more readily motivated extrinsically. Accordingly, it is entirely sensible in this case to select a higher variable component that offers the prospect of a proportionately high reward for commitment. In line with this, Figure 68 (p. 245) shows that, in practice, the variable compensation component is higher on average in product business models and declines increasingly as we move toward complex solutions.

Nevertheless, the level of the variable compensation has to be balanced precisely based on the individual characteristics of the respective business model. Figure 69 can prove helpful in making decisions. The visualization can be used to classify your business models and determine which level of compensation is suitable. As shown in the figure, this is oriented to the sales cycle, the number of transactions, the sales approach, the skills required, the importance of sales to the success of the business model, the variable compensation component compared to the industry as a whole, the autonomy of the sales employees, and the desired time focus. The right half of the figure shows the example of a supplier for car workshops.

Fig. 69 Criteria for the level of the variable compensation

Source: Prof. *Schmitz & Wieseke – Sales Management Consulting*

The following information box shows our research results regarding the optimal variable compensation component in the total compensation.

Information box: Research results regarding the optimal share of variable compensation

Here and at two other points (pp. 274, 343), I would like to deal with a broad-based research project in which we monitored around 50,000 solution projects.

Information about the research project:

The doctoral thesis written at our department by Dr. Felix Brüggemann forms the basis of this project. The results have been published in his dissertation and in double-blind peer-reviewed conference papers.[19]

The 50,000 solution projects were monitored over a period of four years. What was special was that we were able to obtain a 360° view for each project, in which we collected data from the following sources:

- KPIs from the company databases: Including data on solution project sales volumes/profit figures; employee sales volumes/profit figures; employee compensation
- Sales managers: Including data on their management style and manager-to-staff ratio
- Sales teams: Including data on the composition of the sales teams (which actors are involved in the sales process)
- Sales employees: Including data on sales methods and specific behavior data, such as the frequency with which customers were served in individual projects
- Design engineers: Including data on their role in the sales team; behavior data
- Service engineers: Including data on their role in the sales team; behavior data

Research results regarding variable compensation:

We studied the relationship between the variable compensation component and success in the solution business, measured against solution sales volumes.

We found *no linear* relationship between the variable compensation component and the sales employees' success in selling solutions. However, we did determine a *non-linear* relationship (see Figure 70). This relationship can be more accurately described as *inverse U-shaped*. The tip of the inverted U lies at 38%.

Admittedly, this optimum cannot be simply transferred to all industries and companies. The optimal level depends on the respective company's circumstances. Nevertheless, such inverse U-shaped relationships are found time and again. Accordingly, this form appears to be dominant.

---

[19] Brüggemann (2016), departmental doctoral thesis on *Selling Complex Customer Solutions in Industrial Markets – An Empirical Investigation of Success Factors in Sales Force Management and Customer Interactions*.
Wieseke, Alavi, Habel, Schmitz, Brüggemann (2016), *The Ambivalent Role of Variable Compensation in Industrial Servitization*, European Marketing Association Conference, Oslo.
Alavi, Henke, Wieseke, Schmitz, Brüggemann (2019), *The Troublesome Transformation from Product to Service in B2B Contexts: Which Salesperson Behaviors Are effective in Industrial Service Selling?*, American Marketing Association Conference, Austin, TX.

Fig. 70 Risks of variable compensation in the solution business

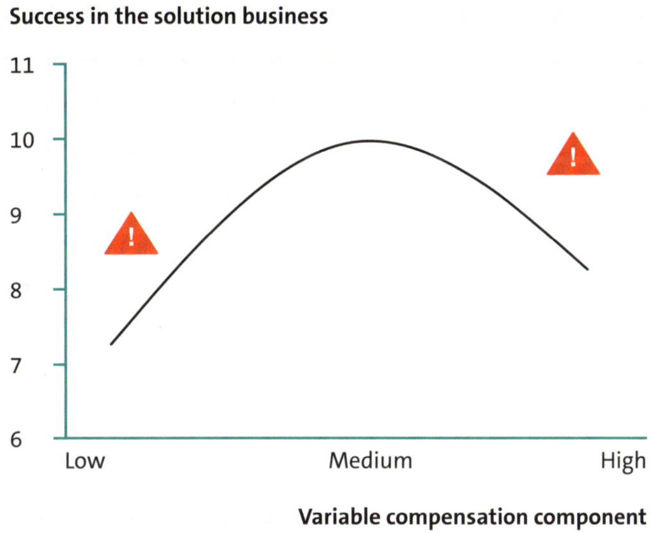

What does this relationship mean to sales management? It means that deviating above or below the optimum reduces the performance of the sales employees in the solution business. And the further we move away from the optimum, the more dramatic the decreases become.

What explanation is there for the inverse U-shaped relationship? In the 50,000 solution projects that we studied, we identified the following two factors based on the data:

- *Work input:* As the variable component increases, the sales employees increase their work input because their compensation is more dependent on it. However, the pressure on the employees to sell as much as possible also increases along with the variable component. As this is precisely the objective of sales employees in the product business, a high variable compensation component should be chosen here.
- *Problem-solving behavior:* Conversely, if the sales employees are put under high pressure to sell in the solution business, they are no longer able to concentrate on finding suitable solutions for their customers' problems. Particularly in the solution business, however, it has been shown that pronounced problem-solving behavior impacts positively on sales success. When deciding on the level of the variable component, therefore, a company should always consider whether it wants to preferably offer an incentive for high selling effort or for pronounced problem-solving behavior with the compensation.

3. Further compensation elements

Apart from the decisions concerning the salary level and the variable compensation component, each compensation system contains additional decision elements. These are concerned with how to handle new recruitments, days of absence, and sales competitions, for example.

For instance, newly recruited sales employees can often be employed in the field salesforce after a short introductory phase in the product business. Consequently, product salespersons are usually also compensated variably from the very beginning. When shifting to the solution business, however, the sales employees should be granted more time for onboarding. As the solutions and sales processes are usually very complex in this case, new employees require more time for training before they can approach customers. The variable compensation should therefore only come into play here when your employees are involved in selling.

The further compensation elements also include sales competitions. These competitions must be harmonized with the objectives of the business model. In the product sector, for instance, the winner should have generated a high sales volume or gross margin, whereas competitions in the solution sector should encourage employees who have successfully sold long-term solution projects. Accordingly, it may be sensible to define competitions with milestones such as the achievement of a specific number of prospective solution customers or the planning of an especially high-value project.

Step 2: Configuring variable compensation

After defining the level of the variable compensation component, attention is turned to its precise structure. Three levers are relevant in this case: Performance measurement, the choice between bonus and commission, and the payment cycle.

4. Performance measurement

The assessment basis for disbursing the variable component is defined in this step. There is a simple recommendation here: The more important the solution business is to your company, the more extensively the variable compensation should be assessed against how successful sales employees are in solution selling. Sales volumes and gross margins of sold services are usually the proven dimensions in this case. Nevertheless, many companies still base their variable compensation components too extensively on product-related sales volumes and gross margins. In such cases, the salesforces are less interested in working toward solution selling.

The further we move in the direction of the solution business, the more long-term and strategic the goals become. Due to the longer sales cycle and the increased consulting effort, your attention should shift from short-term performance indicators to long-term assessment bases such as the cumulative sales volume per customer over the life of a solution contract.

In addition, performance measurement in the product sector should be oriented to the performance of individual sales employees, since a sales pitch is usually conducted one-to-one in the product sector, and sales employees are thus personally responsible for their success. Conversely, responsibility for target achievement in the solution sector is no longer borne by one individual employee as a result of team selling; you should therefore switch to using not only individual performance, but also team performance as the basis of the variable compensation. *DAIKIN* has changed its compensation in this direction (see practical example, p. 259).

5. Choice between bonus and commission

The next decision involves choosing between bonuses and commissions. Commission means that the sales employee receives a percentage of the sales volume or gross margin that is generated, for instance. A bonus is a flat fee granted to the sales employee when specific target levels have been met.

Commissions are often paid in the product business. Even today, there are still companies that pay out more than 50% of total earnings on the basis of commissions. In order to foster high sales volumes in the short term, sales employees are therefore given a percentage of the sales proceeds in the product business. If greater focus is placed on the absolute gross margins, percentage commissions are paid out accordingly.

In the solution business, however, sales income from a specific customer is not only generated at the time of the purchase, but can extend over a longer period of time. The commission logic from the product sector therefore cannot simply be transferred to the solution sector (see information box).

Information box: Switch from commission to bonus – Example of a welding machine manufacturer

A medium-sized manufacturer of welding machines previously motivated its sales employees to sell as many machines as possible by paying commissions. Each time a machine was sold, the employee generated a specific revenue amount for the company and received a defined percentage of this revenue as a commission. To obtain the next commission payment, they had to convince their next customers to make a purchase. As a result, they were continually motivated to sell as many welding machines as possible.

In recent years, new, inexpensive suppliers have increased the competitive pressure on the company, with the result that the welding machine manufacturer took the plunge into the solution business in order to set itself apart from the competition. It now offers not only welding machines, but also a range of additional services. Moreover, the machines are now also leased. For this, the customers do not pay a high one-off price for the machine, but small monthly sums throughout the entire contract period. If the company were to continue paying its sales employees a commission for the generated revenue, they would no longer receive a high reward following the sale, but small amounts each month instead. This would demotivate the sales employees because these amounts would be low. The management therefore decided to switch to bonus payments and to pay their sales employees a higher reward immediately after contract completion.

Figure 71 takes seven parameters into consideration (sales cycle, sales as a causal success driver, team (vs. individual) as a success factor, relevance of non-sales activities, comparability of sales areas, predictability of demand, and target-setting quality). It offers a decision-making aid for determining whether bonus or commission payments are more sensible for your company. The aforementioned supplier for car workshops is again shown as an example in the right half of the figure.

6. Payment cycle

Another compensation model decision area is the payment cycle. This defines how frequently the variable compensation is to be paid out to the sales employees. The payment cycle should essentially be oriented to the sales cycle. This means that the payment cycles should be shorter in the product sector and longer in the solution sector. Since contract completions take longer in the solution business than in the product business, the sales employees should accordingly receive a bonus less frequently.

Fig. 71 Criteria for choosing between bonus and commission

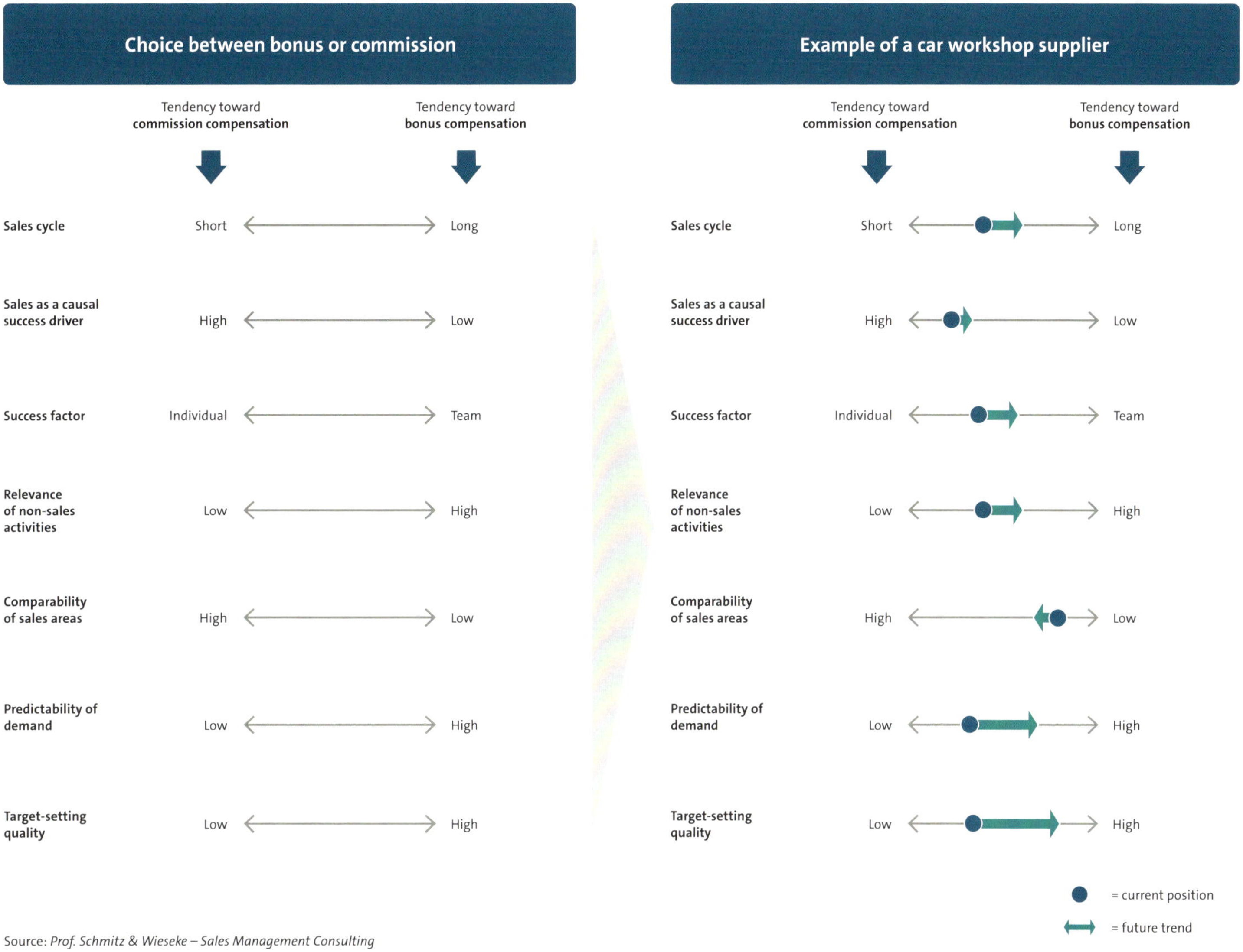

Source: *Prof. Schmitz & Wieseke – Sales Management Consulting*

Step 3: Defining the payout curve

The third step for structuring the compensation models involves defining the payout curve. The principle of the payout curve can best be explained on the basis of Figure 72 in the following. It shows the level of the salary payment as the target achievement level progresses. As can be seen, the salary is subdivided into three zones: 1) Fixed salary, 2) variable salary, and 3) over-performance salary.

Fig. 72 Parameters for determining the payout curve

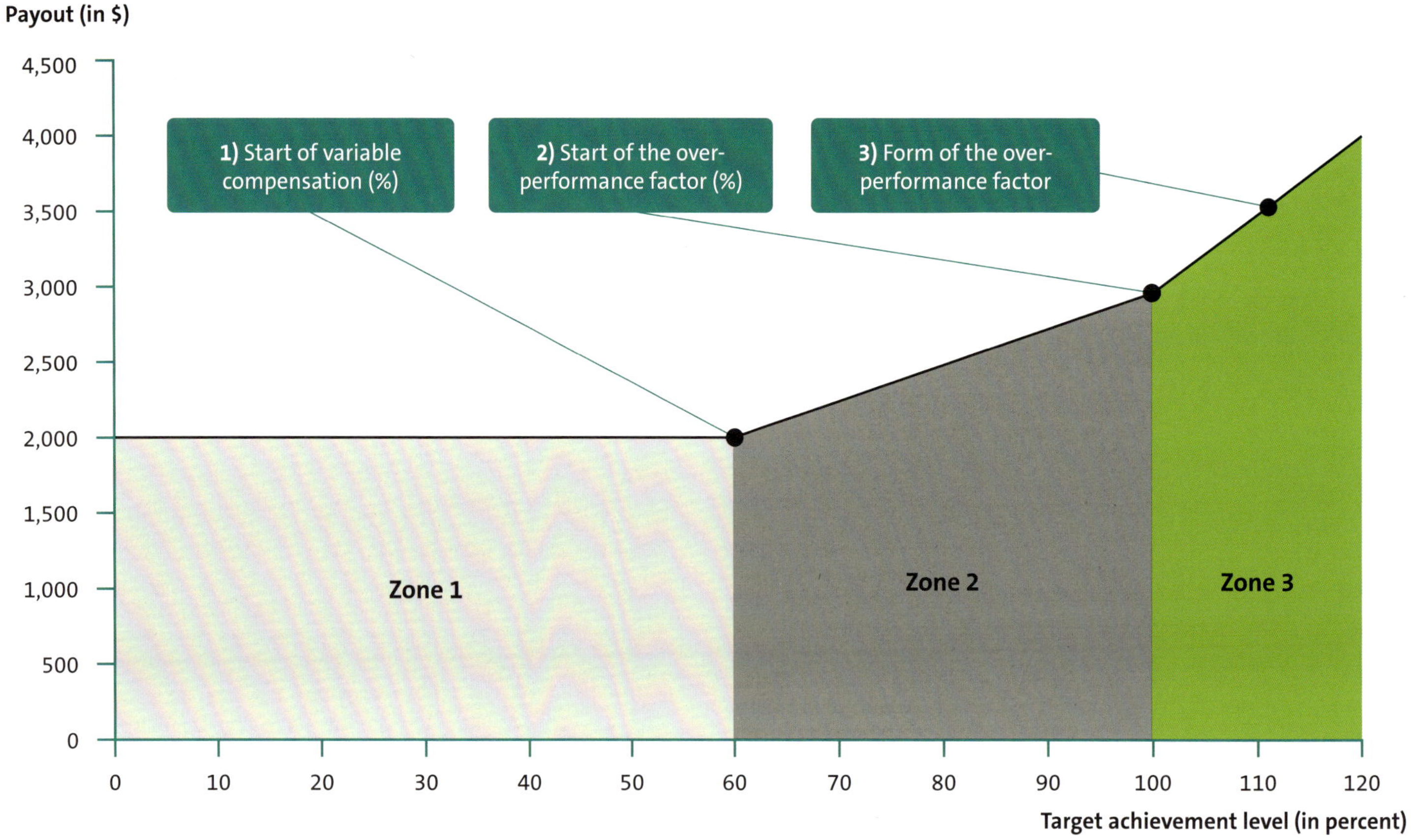

Three management decisions have to be made here:

- At which target achievement level does the variable compensation start?
- Is there an over-performance factor at a specific target achievement level?
- Is the over-performance factor used progressively or degressively?

7. Start of variable compensation

The first controlling lever for defining the payout curve is the start of the variable compensation. In the example shown, the fixed salary is \$2,000 (zone 1). Payout of the variable component of the compensation starts at a target achievement level of 60%. At 100% target achievement (end of zone 2), the variable salary is \$1,000.

The point at which the variable compensation starts depends on how well the sales targets for the respective business model can be specified. Accordingly, a general recommendation is difficult here. The following rule of thumb applies: If a task is still new and unfamiliar to sales employees, even small steps on the road to success should be rewarded. This lowers the hurdle of treading new ground.

8. Start of the over-performance factor

The next decision concerns the question of whether there is to be an over-performance factor and, if yes, at which target achievement level. This involves the issue of how to deal with above-average performance results. A high over-performance factor is intended to motivate sales employees to give their best even after achieving the specified target. This is shown in zone 3 in the figure above: In the example, the over-performance factor starts at 100%.

9. Form of the over-performance factor

A *progressive* over-performance factor means a steeper increase, or a "boost," in the variable compensation at the defined target achievement level. Conversely, a *degressive* over-performance factor means that the variable compensation increases less steeply, that is, it is dampened or completely capped at the defined target achievement level. In the example in the figure above, the upslope in the over-performance area (zone 3) is steeper than before, meaning that it is progressive.

Handling the over-performance factor remains an individual decision on each business model step. However, high over-performance factors tend to be seen more frequently in the product business.

To visualize the relationships between the individual controlling levers of the payout curve, we have developed a **payout simulator** (see information box). It can be used to create various payout scenarios and therefore see how the curve changes each time the individual parameters are adapted.

Information box: The *payout simulator* for new compensation models

The *payout simulator* can be used to determine the effect of a compensation switch in two directions.

a) Effect on your operating results

b) Effect on the individual compensation of each sales employee

Both aspects should be clear prior to compensation switches. Three steps can be identified in this process.

Step 1: Definition of the curve parameters and modeling of the curve

The following parameters must be defined in advance to be able to calculate the payout curve:

1. Variable compensation component (50% in the example shown in green in Figure 73, 30% in the example shown in blue)
2. Target achievement level at which payout of the variable compensation begins (50% in both examples)
3. Target achievement level at which an over-performance factor comes into play (100% in both examples)
4. Gradient in the over-performance area (2.0 in both examples)
5. "Cap threshold," i.e., a boundary at which a further increase in the variable compensation is throttled (120% in both examples)
6. Gradient after the cap threshold (0.3 in both examples)

This data can be used to determine what percentage of the scheduled salary is to be paid out at each target achievement level.

The *payout simulator* can also be used to model two (or several) different curves. In the example in Figure 73, a company decided to provide its employees with a choice of two payout curves: The performance model (green curve) involves higher risks because a lower fixed-compensation component is paid out. However, it is more attractive for employees who surpass their targets, as more is paid here than in the standard model.

Fig. 73 Example of a payout simulator

Payout simulator

Parameters of the standard model (blue line)

1)	Variable compensation component in total compensation	**30%**	4)	Gradient in the over-performance area (here 100–120%)*	**2.0**
2)	Start of variable compensation (in % of target achievement)	**50%**	5)	Cap threshold (in % of target achievement)	**120%**
3)	Start of the over-performance area (in % of target achievement)	**100%**	6)	Gradient after the cap threshold (here as of 120%)*	**0.3**

Parameters of the performance model (green line)

1)	Variable compensation component in total compensation	**50%**	4)	Gradient in the over-performance area (here 100–120%)*	**2.0**
2)	Start of variable compensation (in % of target achievement)	**50%**	5)	Cap threshold (in % of target achievement)	**120%**
3)	Start of the over-performance area (in % of target achievement)	**100%**	6)	Gradient after the cap threshold (here as of 120%)*	**0.3**

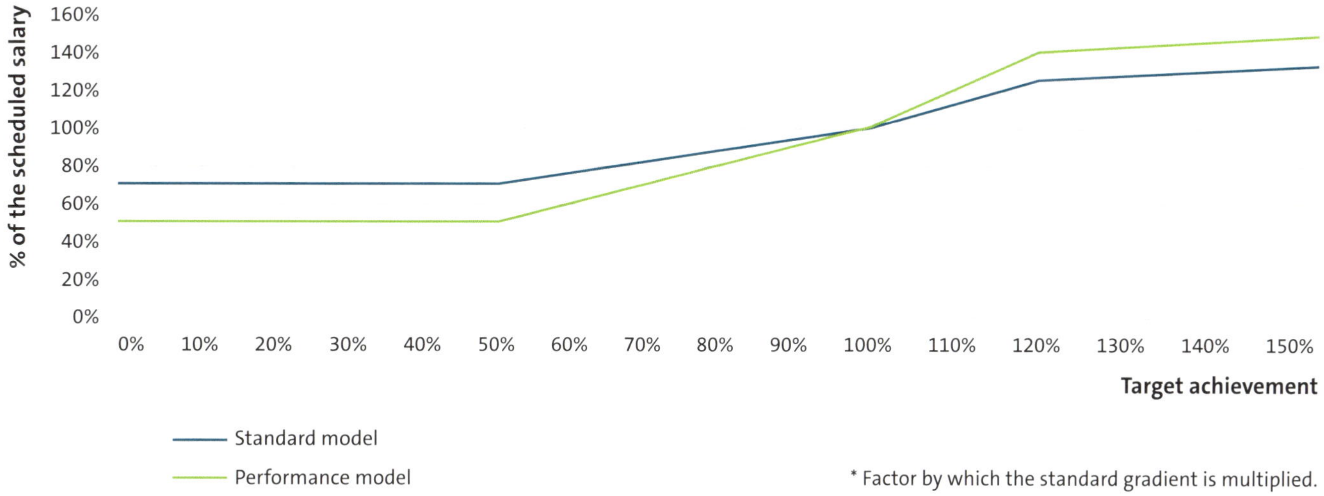

* Factor by which the standard gradient is multiplied.

Source: *Prof. Schmitz & Wieseke – Sales Management Consulting*

SALES

Step 2: Simulation of the new compensation model using real employee data

In the next step, information has to be collected about the employees who are to be provided with the new compensation model. The following parameters must be defined in this case:

- Scheduled salary (i.e., the salary that would be paid out at 100% target achievement)
- Target achievement levels for the success variables on which the new compensation model is to be based. Either values referring to the past or forecast values can be used for this. The weighting of the targets should be taken into consideration (e.g., 80% of the variable compensation for target 1 and 20% for target 2).

The analysis period can be monthly or annually, whereby the annual analysis utilizes a more representative period of time due to seasonal fluctuations. It is important that the scheduled salary and the target achievement levels each refer to the same analysis period (i.e., monthly or annually).

This information can be used to simulate which gross salary each employee would receive as part of the new compensation model with reference to one month or one year, for example: The percentage of the scheduled salary that an employee would receive for their personal target achievement level is determined by the payout curve, and then this percentage value is multiplied by the employee's actual scheduled salary.

These analyses can be significantly speeded up by using the *payout simulator*. This is especially helpful if several targets, possibly with different weightings, are additionally factored in.

It may also be the case that individual employees have different variable compensation components due to historically evolved employment contracts, and that these cannot possibly be changed without a great deal of effort. A customized curve can then be developed for each employee: The system would remain the same for all employees, but different scheduled salaries and variable components would be used as the basis for the calculation.

Step 3: Useful additional analyses

Once the new gross salaries have been simulated, a range of useful additional analyses can be performed and used to assess the new compensation model:

- Effects on the company's total costs:
  - How does the gross salary sum change across all of the affected employees?
- What are the cost impacts if a 12-month "period of grace" is defined to introduce the new compensation model (i.e., if over-performance is rewarded according to the new model, but under-performance is not penalized accordingly)?
- Comparison of alternative compensation models
- Winner/loser balance sheet (see Figure 74):
  - How do the new gross salaries differ from the gross salaries previously paid out?
  - Which employees fare better, and which fare less well? How high are the absolute and percentage changes?

Fig. 74 Example of a winner/loser calculation

Employee	Scheduled salary with 100% target achievement	Variable component	Fixed component of the salary	Variable component of the salary	Assumed target achievement	Salary as per old compensation regulation	Salary as per new compensation regulation	Δ (absolute)	Δ (%)
A	40,000.00	20%	32,000.00	8,000.00	93%	40,000.00	39,440.00	−560.00	−1.4%
B	45,000.00	20%	36,000.00	9,000.00	105%	45,000.00	45,450.00	450.00	1.0%
C	50,000.00	20%	40,000.00	10,000.00	97%	50,000.00	49,700.00	−300.00	−0.6%
						135,000.00	134,590.00		

Let us take a look at a more detailed practical example of adapting compensation models on the road to solution business models (see information box).

Information box: The development of a new compensation model for *DAIKIN*

DAIKIN is the world's leading manufacturer of heat pumps as well as ventilation and climate control systems. The company operates in the private, trade, and industrial customer business. It was founded in 1924 in Osaka, Japan, where its headquarters are still located. *DAIKIN Airconditioning Germany GmbH* is *DAIKIN's* German sales subsidiary. It has a staff of around 180 employees and generates annual sales of around $164 million. Within Germany, *DAIKIN Airconditioning Germany GmbH* (called *DAIKIN* in the following for the sake of simplicity) is subdivided into six regional offices.

DAIKIN's value chain is relatively complex and requires various sales roles. Air conditioning systems are ultimately bought and installed by plant manufacturers (see Figure 75, constellation #1). However, some of the projects that the plant manufacturers implement are initially planned by engineering offices (so-called planning offices, see Figure 75, constellation #2) and then passed on to the plant manufacturers with very concrete specifications regarding the air conditioning system manufacturer. In some cases, particularly in major projects, the manufacturer's requirements are specified to the planning offices by the investor (see Figure 75, constellation #3). *DAIKIN's* target groups accordingly consist of investors, planning offices, and plant manufacturers. One or more sales roles are necessary for each target group in order to ensure that each is addressed perfectly.

The three central roles have different task profiles. Each of these roles involves separate solution selling requirements. As *DAIKIN* is currently transforming from a product to a solution supplier, these requirements are increasing considerably.

- Consulting sales (CS): CS salespersons support planning offices in project planning and attempt to influence the customers in such a way that they plan *DAIKIN* products. The CS salespersons are usually engineers.
- Dealer sales (DS): DS salespersons address plant manufacturers during the implementation of projects specified by planning offices or during the planning of smaller projects. The DS salespersons are usually refrigeration engineers.
- Specialist consultants (SC): Specialist consultants are specialists in various technologies (e.g., cold water systems, heat pumps, or commercial refrigeration). They are called on as required and mainly address the plant manufacturers.

Fig. 75 Sales constellations at *DAIKIN*

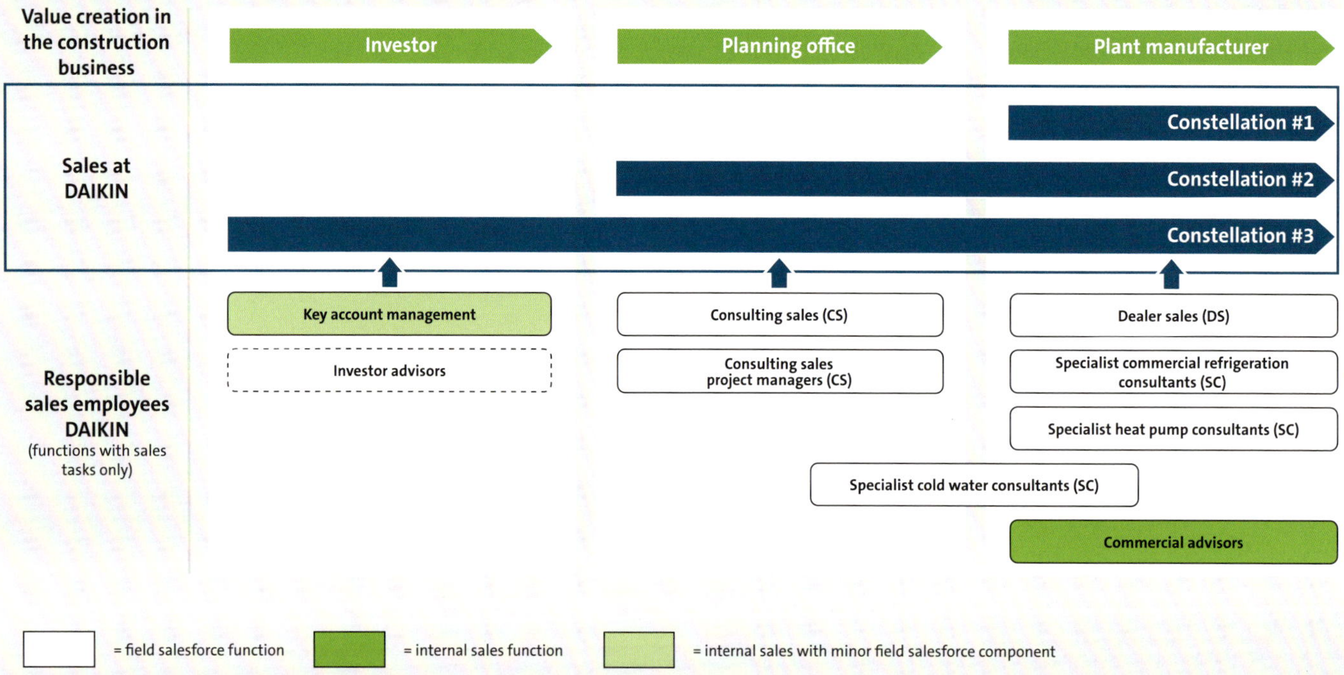

Step 1: Understanding of the status quo

An understanding and the integration of two perspectives are important to be able to develop an effective and satisfactory compensation system:

- Company perspective: What is the current strategy or the current business model? How should the strategy/business model change over the next three to five years?
- Employee perspective: Which needs of the employee predominate in different sales roles? Where do barriers to solution selling need to be broken down?

In order to precisely understand these perspectives, we conducted ride-alongs with employees in various sales functions, workshops with employees and managers, and a standardized survey. The intensive involvement of the employees in developing the new compensation system subsequently led to a high level of acceptance of the new model.

The analysis of the status quo revealed the central weaknesses of the previous compensation model. Interface management and project tracking proved particularly difficult: Ideally, planned projects should be transferred from the CS salespersons to the DS salespersons to enable consistent processing. Due to the previous misplaced incentives in the two groups' compensation, this cooperation had also failed to function adequately thus far.

The enormous administrative effort in determining the variable compensation was additionally classified as problematic. So far, CS salespersons had been compensated based on the sales volume that arose due to projects being planned with the planning offices. However, proper tracking of the projects did not take place and could not be mapped using the sales systems at *DAIKIN*. This means that it was often unclear whether projects that were actually built and had generated sales had originally been planned by the CS salespersons. Consequently, the CS salespersons attempted to implement their own project tracking using self-constructed Excel tools. This led to intensive discussions with the personnel department when the sales management transferred completely different values than those determined by the respective CS salespersons themselves for calculating the annual commission.

Finally, *DAIKIN* decided to freeze all sales employees' compensation due to the multitude of flaws in its previous compensation system. For three years, the employees therefore received a salary that was set at a fictitious target achievement of 100%: Paradise for the low performers, who were delighted to receive a salary hike, but yet another frustration factor for the high performers, some of whom were then earning far less than what they were capable of.

Step 2: Development of a new compensation system (nine controlling levers)

In the second step, we developed a new compensation model in the context of several workshops with the *DAIKIN* management. To do this, we worked through the nine controlling levers of our concept (Figure 76).

Fig. 76 The nine controlling levers for assessing variable compensation

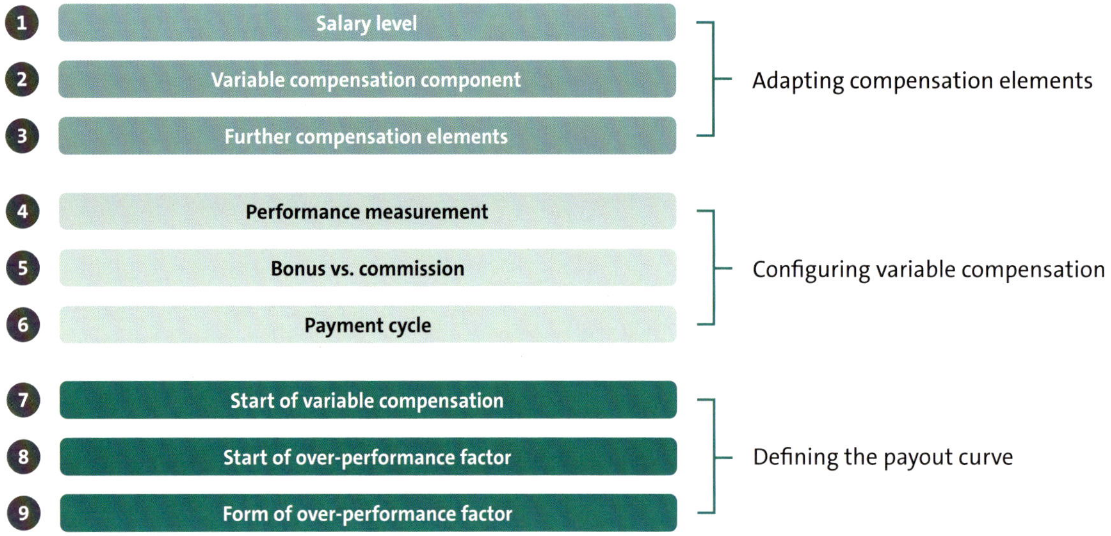

Source: *Prof. Schmitz & Wieseke – Sales Management Consulting*

This led to a new compensation model; I deal briefly with its characteristics in the following. Figure 77 provides an overview of the basic data.

Team and individual targets

The inclusion of team targets was central to the development of the new compensation model. Within this, a significant part of the variable compensation for all sales functions was to be derived from the sales generated in the regional office.

In turn, the prerequisite for high regional office sales is interfunctional cooperation between the sales roles. The demands on managers increased as a result. This was addressed in the context of training courses and discussions with the managers. Vertical integration of the targets was also crucial, that is, the targets were harmonized across hierarchy levels as well. Therefore, in the new compensation model, the regional sales manager (head of the regional office) is also compensated based on target achievement in the regional office. In the case of nationwide cooperation, the regional sales managers are additionally compensated on the basis of national target-setting.

Fig. 77 The new compensation model of *DAIKIN*

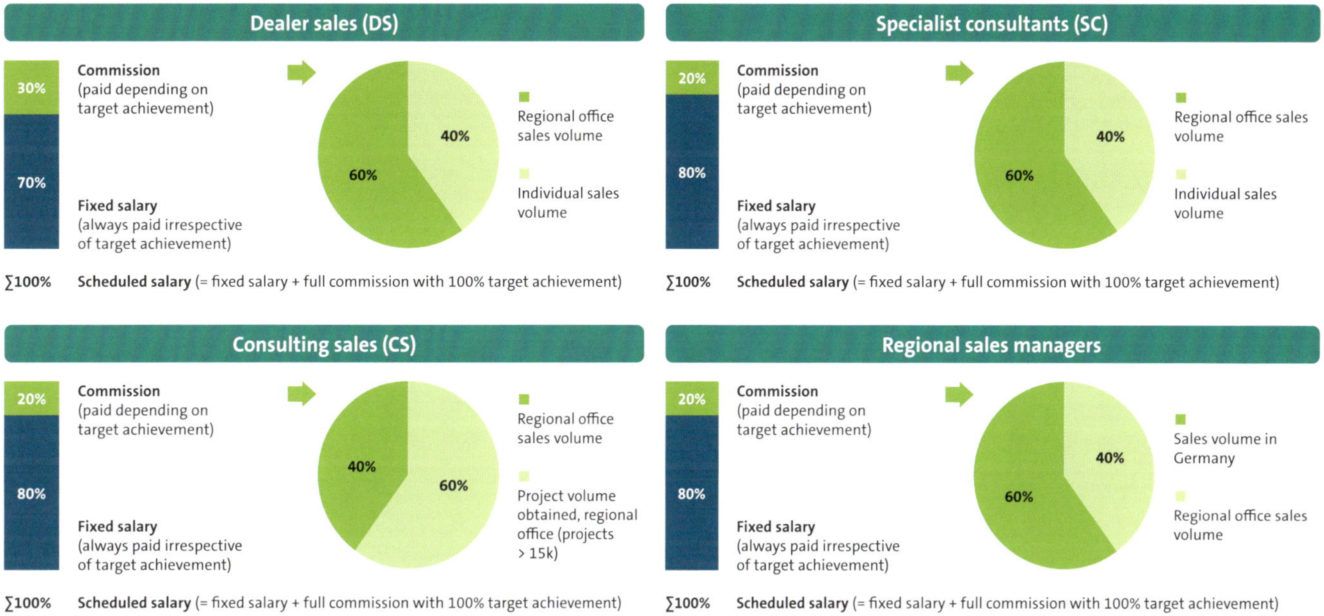

Step 3: Evaluation of the new model

We undertook a comprehensive risk analysis to evaluate the new model. Besides qualitative change risks (e.g., increasing fluctuation due to a lack of acceptance of the new model), this also included quantitative business risks. In particular, the effects of the new model on the total salary compared to the current compensation system and the effect on individual employees were analyzed here (winners/losers of the new system, see p. 259).

Step 4: Introduction of the new model

Workshops were conducted for the employees in order to communicate the new model. A great deal of time was scheduled (one day per sales function) for each role – including on the part of the *DAIKIN* management – so that the employees felt fully informed and secure upon introduction of the new model. The employees were also provided with an FAQ list, an explanatory document, and an Excel-based calculator to enable them to monitor their target achievement and their forecast payout throughout the year.

SALES – IMPLEMENTING NEW BUSINESS MODELS IN SALES

Chapter 11
Adapting sales leadership

Project steps

- Step 1: Setting sales targets
- Step 2: Guiding sales employees
- Step 3: Assessing sales employees

Message of this chapter

Sales leadership has to be adapted for each new solution business model.

What should you have learned by the end of the chapter?

- You are familiar with the three project steps for adapting sales leadership for your solution business models.
- You are able to define the adequate leadership profile for first-line sales managers for each business model step.

DEATH VALLEY MARKETS

Strategy of minimum costs and prices

GREEN HILL MARKETS

Strategy of solution business models

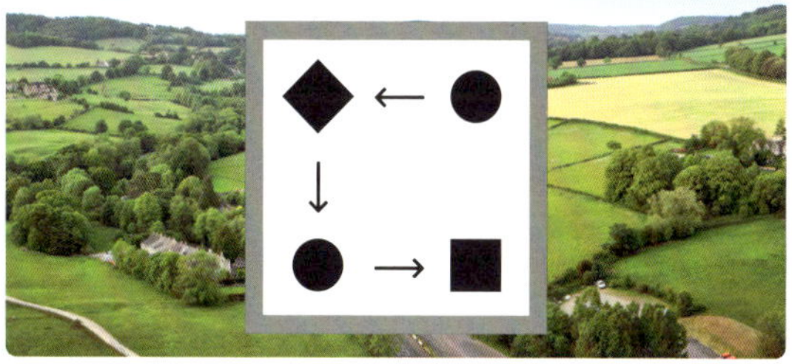

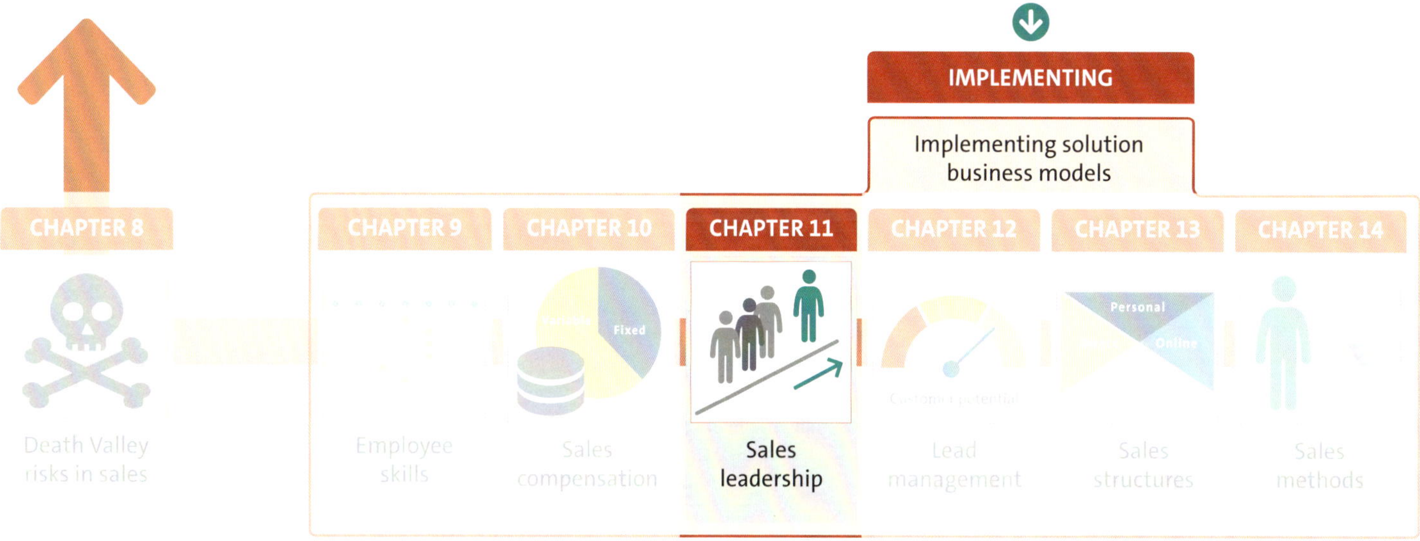

SALES

This chapter deals with the requirements of sales leadership. Three steps are necessary to adapt it to the requirements of new business models (see Figure 78).

Fig. 78 Implementing new business models – project steps in sales management

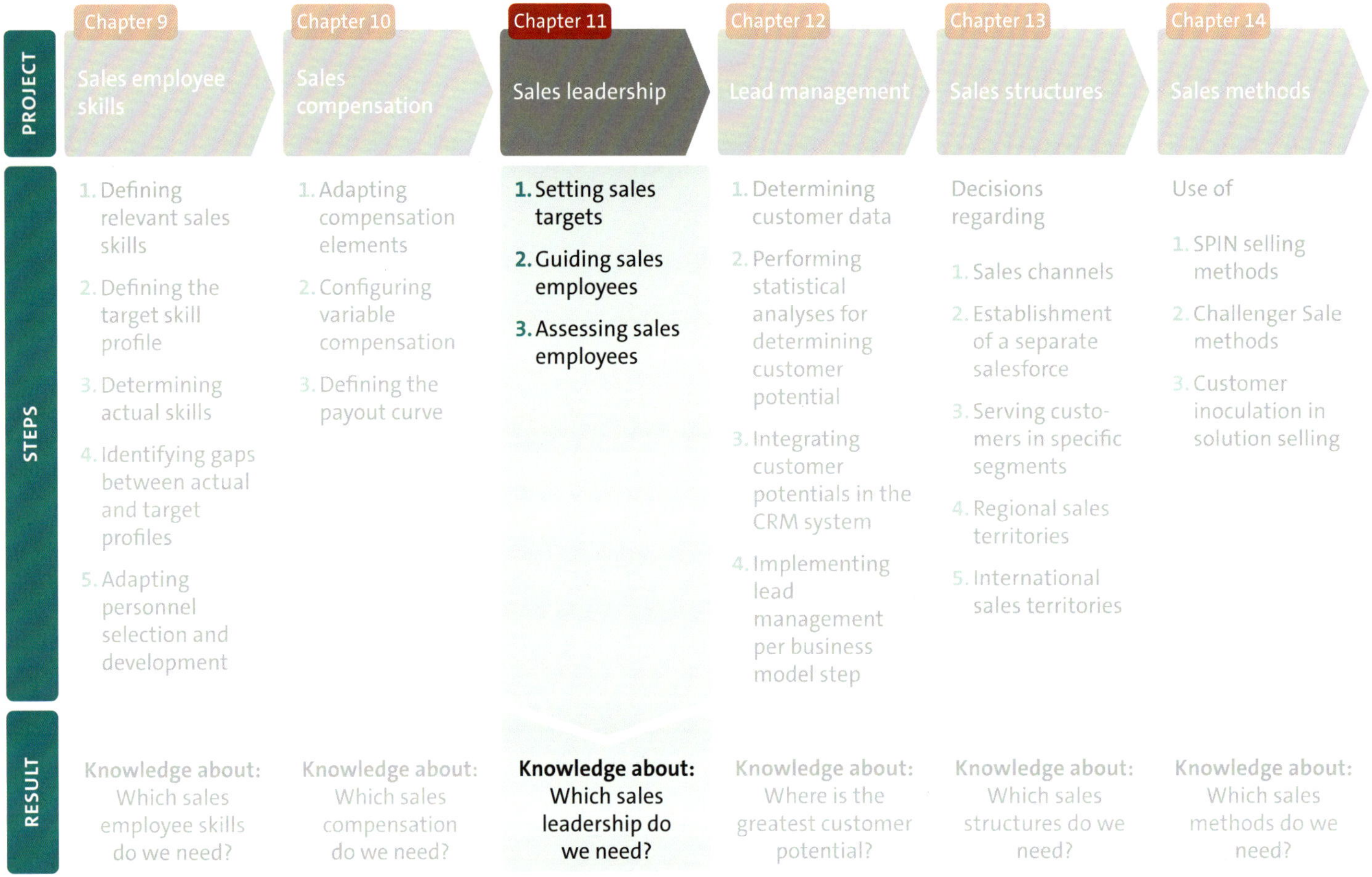

"The switch to solution selling marks a dramatic change for our sales employees. They need more support from their supervisors than ever before." Sales manager of a manufacturer of large machines in the construction industry

The requirements for sales leadership also change when shifting to new solution business model steps. We can determine how they differ in the individual business models on the basis of 10 criteria (see Figure 79). These can be assigned to three higher-level project steps: 1) Setting sales targets, 2) guiding sales employees, and 3) assessing sales employees.

Fig. 79 Leadership profiles depending on business model steps

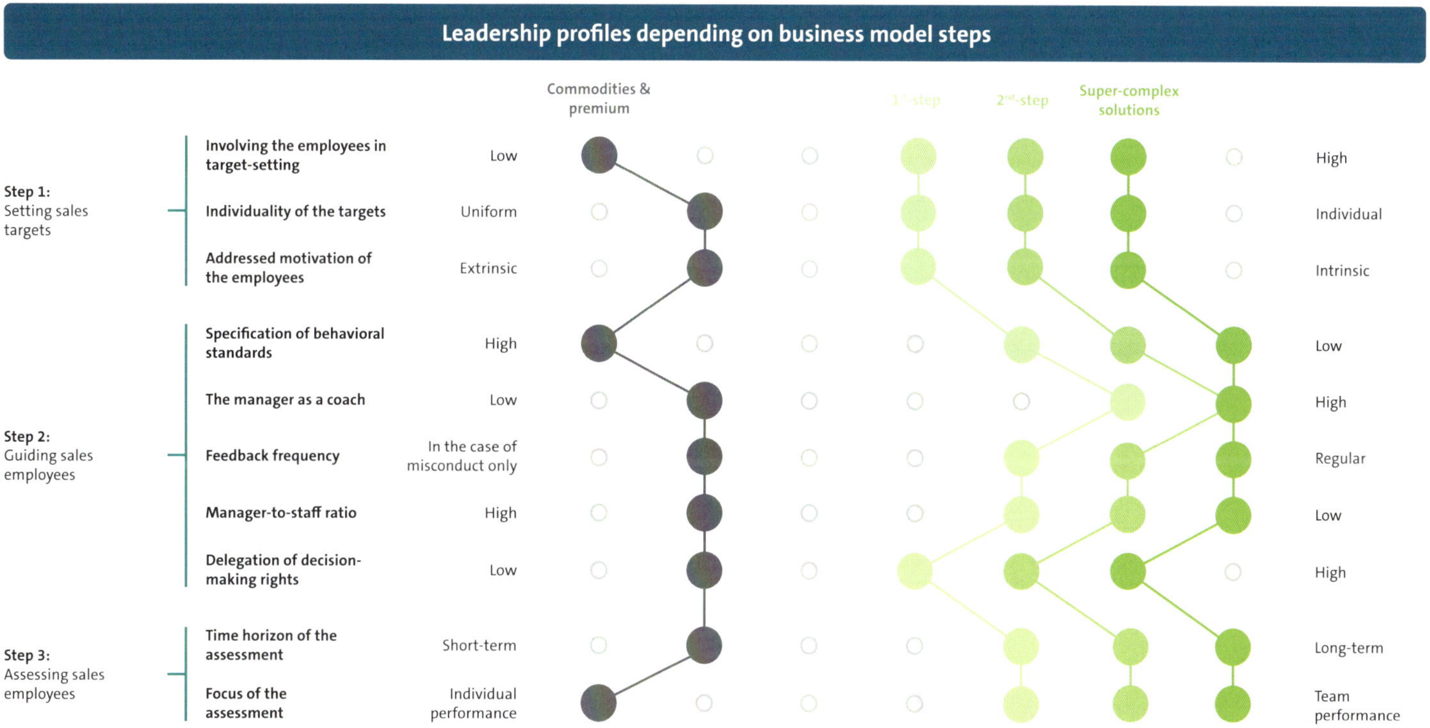

The discrepancy between the requirement profiles is shown in Figure 79. It shows the average profiles of product business models (commodity and premium) compared to those for the three solution business model steps (1st-step, 2nd-step, and super-complex solutions).

Of course, the ideal values in the 10 leadership criteria differ from company to company. I therefore describe our average observations in the following. The focus here is on leadership by *first-line sales managers*, meaning the sales employees' direct supervisors.

Project steps

Step 1: Setting sales targets

When entering a new business model step, new corporate goals – and therefore new sales targets – are also introduced. The role of sales leadership involves breaking the corporate goals down to the level of the sales employees. Three decision areas are important in this process: Involving the employees in target-setting, the individuality of the targets, and the addressed motivation of the employees.

Involving the employees in target-setting

How extensively the managers take the employees' assessment into account in defining targets is defined here. In product business models, the sales targets are usually specified by the senior management levels. The sales managers then break these targets down to the individual employees. The sales employees are rarely involved in target-setting.

A different procedure should be chosen in solution selling. Here, decisions are more strategic in nature and encompass more diverse areas than in the product business. Due to the increased complexity, it is entirely sensible to involve the sales employees and their individual capabilities more extensively here in target-setting and to jointly develop milestones. However, ultimate decision-making authority concerning the distribution of the targets to the individual employees should still lie with the sales manager.

Individuality of the targets

The individuality of the targets describes the extent to which specific or uniform targets are set for all employees. The same targets primarily apply to all employees when selling products. Of course, you can also include individual components such as the respective sales area or the type of product. Essentially, however, the rule is that two employees in the same position means the same target criteria apply.

Due to the more intensive strategic focus and the more diverse tasks involved, however, this rule cannot simply be transferred to the solution business. In this case, you should instead switch to distributing the targets and tasks as per the employees' respective capabilities.

Addressed motivation of the employees

The type of target-setting can address either the employees' intrinsic motivation or their extrinsic motivation. Due to the uniform and clear targets in the product business, intrinsic motivation plays a comparatively minor role there. However, this often means that the sales employees only take on additional work when they are also accordingly rewarded for it. This type of incentivization therefore primarily addresses the sales employees' extrinsic motivation.

Conversely, sales employees in the solution business should be motivated more to contribute to the overall success of the team by fulfilling their individual tasks. As a result, all employees pursue a common goal, leading to the emergence of a sense of togetherness that above all motivates the sales employees intrinsically. Excessive predefinition of the targets in this case hampers the motivation to look at the big picture.

Information box: Is the classic "SMART" formula enough?

In our work with companies, we are often asked if the classic "SMART" formula is sufficient to set sales targets for solution business models. SMART stands for targets that are *Specific*, *Measurable*, *Attainable*, *Realistic*, and *Timely*.

Figure 80 applies the SMART criteria to the five business model steps. It shows that some of the criteria cannot be transferred from the product business to the solution business models. It is therefore considerably more difficult in the case of complex solutions to derive specific targets that are measurable in detail and to check them at short time intervals.

We therefore recommend a "SMART+" approach that adds the criteria (1) *individualized*, (2) *choice-based*, and (3) *all-or-nothing*. *Individualized* and *choice-based* pick up on the above descriptions and are aimed at establishing the highest possible level of self-motivation. *All-or-nothing* means that the sales cycles of individual orders are so long – especially in the case of 2nd-step solutions and super-complex solutions – that a customer's individual commitment or rejection is equivalent to 100% *(all)* or 0% *(nothing)* target achievement. If the sales team actively factors this into target-setting, it fosters an entrepreneurial mindset and self-organization within the team.

Fig. 80 "SMART+" as the standard for target-setting in solution business models

		Commodities	Premium products	1st-step solutions	2nd-step solutions	SCS
S	**S**pecific	"Easy tasks" – more granular targets conceivable and sensible (e.g., number of visits, number of new customers → sales volume, gross margin)			"Complex tasks" – high-level targets more sensible (sales volume, gross margin, team targets)	
M	**M**easurable		Usually easily measurable		Behavior that leads to more sales volume is difficult to measure (e.g., interaction quality, team/networking skills)	
A	**A**ttainable	Always dependent on the defined target level and the respective area/customer potentials				
R	**R**ealistic	Always dependent on the defined target level and the respective area/customer potentials				
T	**T**imely	1 month	...	...	...	≥ 1 year
		Ever-longer sales cycles necessitate longer target-setting periods				
+	Individualized	Necessary in the case of little standardization				
	Choice-based	Fosters self-organization and entrepreneurship				
	All-or-nothing	Fosters self-organization and entrepreneurship				

Conclusion
- SMART is not sufficient in the case of solution business models
- A more complex target-setting process is necessary

Step 2: Guiding sales employees

In the next step, you should define how to encourage your employees to behave in keeping with the strategic business model goals. Again, there are five decision areas here that you should include in your selection: Specification of behavioral standards, the manager as a coach, feedback frequency, manager-to-staff ratio, and delegation of decision-making rights.

Specification of behavioral standards

The specification of behavioral standards in sales leadership shows how precisely sales managers should dictate to their employees how they should behave. In the product business, very tight management is often implemented, in which first-line managers specify a clear framework within which their salesforce is able to operate. Among other aspects, this includes the number of sales calls that sales employees should conduct within a week and guidelines for sales pitches. The employees are measured against whether they observe these specifications or not.

In solution selling, the solutions are adapted specifically to the respective customer. The sales employees should therefore be given more leeway here with regard to their behavior. Only then can they deal with each individual customer optimally. Accordingly, it is better to set down fewer behavioral standards for sales employees in the solution business.

Another aspect is the development of what are called "sales playbooks," which serve to support the salesforce and are not intended to act as restrictive behavioral standards. They exist in many companies for the "Challenger Sale" method, which I deal with in Chapter 14.

The manager as a coach

Sales managers can act as controllers or coaches for their employees. In product selling, the manager will check whether the sales employees behave according to the specified standards and achieve their targets more extensively than in the solution business.

Due to the higher level of complexity and more individual tasks, such a procedure can jeopardize the intrinsic motivation of sales employees in the solution business. Here, managers should instead support their employees much more extensively through individual improvement suggestions. Many sales employees require close coaching, particularly while finding their feet in the unaccustomed solution business (see information box on the uncertainty of sales employees in the solution business). Because of this, companies readily use additional field coaches during the initial phase.

Feedback frequency

Managers can provide their sales employees with feedback either on a regular basis or only when the situation necessitates it. In the product business, sales employees often know precisely how they can achieve their targets due to the tight behavioral standard specifications. Managers therefore offer less feedback here by comparison. The manager only intervenes when the employee deviates from the behavioral standards.

In the solution business, the sales method is less standardized, so the sales employees are reliant on finding out whether they are meeting expectations or where they can improve. Sales employees in the solution sector should be supported with regular feedback. Individual recommendations and improvement suggestions help them to continue developing.

Information box: How can the uncertainty of sales employees be managed in the solution business?

The solution business not only means uncertainty for customers, but also for sales employees. When progressing from product to solution selling, the sales employees have to leave their comfort zones and tread new ground. In addition, their success is now dependent on others more than ever before. This means that they have to relinquish control to some extent. Their level of control decreases the more complex business becomes and the more the sales employee has to rely on other people involved with the sales team and the buying center.

According to our findings, too much uncertainty can negatively impact sales success. This uncertainty leads to the sales employees' becoming less effective in recognizing customer needs and in scrutinizing customer business models. The underlying psychological process is easy to understand: Sales employees suffering from uncertainty are unable to concentrate sufficiently on the complex customer analysis.

In his doctoral thesis, Dr. Felix Brüggemann examined the effect of three classic sales leadership tools on the perceived uncertainty and the performance of sales employees in solution selling (see Figure 81).

Fig. 81 Sales leadership approaches for dealing with sales employee uncertainty in the solution business

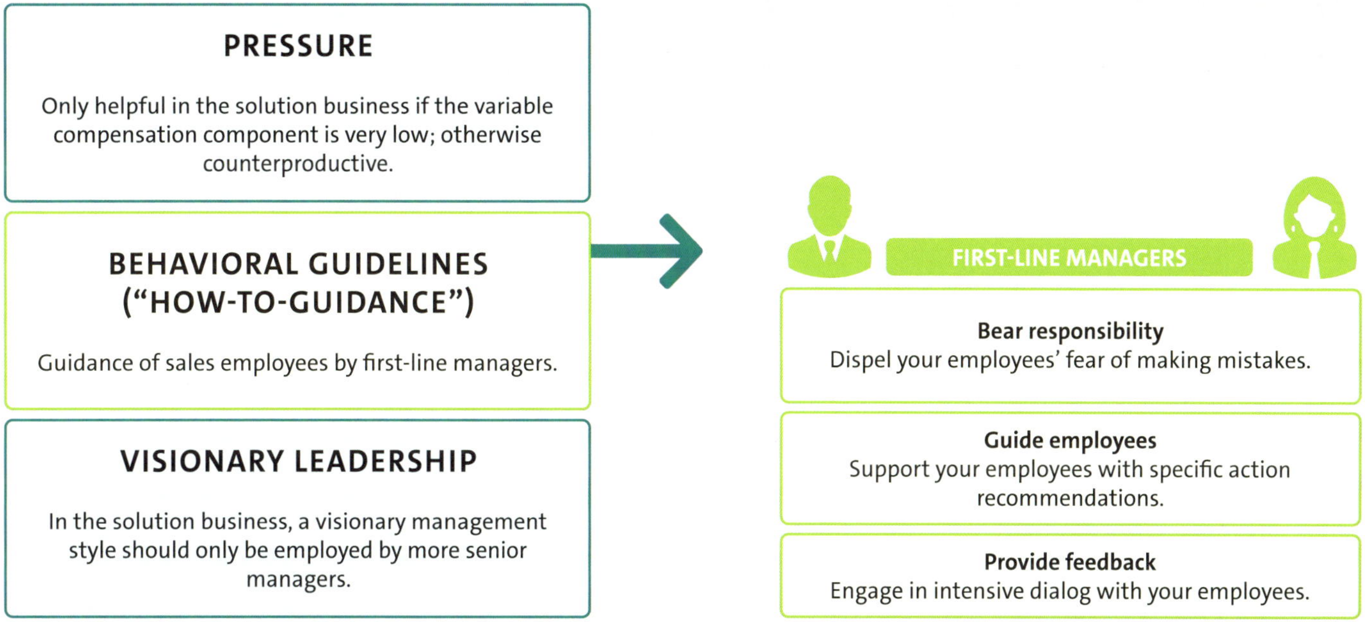

Source: Doctoral thesis, Dr. Felix Brüggemann, Sales Management Department, Ruhr University of Bochum

According to his thesis findings, there are very few exceptions when it is not counterproductive for supervisors to increase pressure to sell in the solution business. Often, they also attempt to motivate employees with a positive "vision of a successful solution supplier." However, visionary leadership increases the level of uncertainty of many sales employees who are new to solution selling if the manager neglects the individual steps on the path to this vision or does not communicate them correctly.

Conversely, it is positive if first-line sales managers provide their employees with behavioral guidelines ("how-to guidance") that reduce their level of uncertainty. In addition, they then have more capacity to attempt to understand customer business models. With his data, Dr. Brüggemann was able to verify the causal chain from increased "how-to guidance" to increased problem-solving at the customer's site and improved solution performance on the part of the sales employees.

Manager-to-staff ratio

The manager-to-staff ratio describes the number of sales employees managed per manager. Here, a high manager-to-staff ratio goes hand-in-hand with a reduced level of coaching intensity per employee, and a lower manager-to-staff ratio with a higher level of coaching intensity. In product selling, sales employees are usually provided with precise specifications in the form of behavioral standards and clearly defined targets. This enables first-line managers to verify how well individual sales employees are performing and the points at which they themselves need to intervene. Due to the high level of transparency, a higher manager-to-staff ratio should be used in the product business.

However, a high manager-to-staff ratio in the solution business would mean that first-line managers would be unable to fulfill the recommended close coaching of their employees and provide the necessary feedback. In solution selling, it should therefore be ensured that the number of sales employees assigned to first-line managers is such that each and every employee can be dealt with individually. Accordingly, a lower manager-to-staff ratio should be selected in the solution business. When entering 1st-step solutions, for example, no more than six sales employees per first-line manager are recommended.[20]

Delegation of decision-making rights

The delegation of decision-making rights means the extent to which sales employees can make decisions themselves. The most important areas are:

- The structure of the sales process (the number of customer visits, for example)
- Pricing (discounts, for example)
- The configuration of a product or a solution
- The structure of service contracts (the contractual period, for example)

In product-centric business models, such decisions are more likely to be defined centrally. The background in this case is that the sales employees should structure the sales pitches with individual customers as efficiently as possible, and that customer-related adaptation of products would equate to an additional time requirement. Excessive price sovereignty in the product business also frequently tempts employees to choose the “simple” route and grant price reductions too often.

However, customer-related adaptations are sensible in the solution business. Here, customers want individual solutions, which the sales employees are only able to develop in discussions with them. Where serving customers and configuring solutions are concerned, you should therefore grant your sales employees more decision-making rights than in the product sector.

[20] According to a study conducted by my colleague Christian Schmitz, the average manager-to-staff ratio of first-line sales managers in sales across all business model steps is 1:8.

Step 3: Assessing sales employees

Depending on the specified targets, different standards can be used to assess the employees. We will therefore look at two decision areas for assessing sales employees in the third step: The time horizon of the assessment and the focus of the assessment.

Time horizon of the assessment

A manager can apply either a short-term or a long-term perspective for the employee assessment. If the employees can achieve their targets in the short term, their assessment should also take place after a short time. With a very few exceptions (in the case of complex machines, for example), this is the case in the product business. Work here is often carried out using monthly or quarterly sales targets.

In contrast, the targets in the solution business are more strategic and therefore long-term. Since solutions are much more complex and require more explanation than products, they also necessitate longer consultations. This means that sales employees are able to sell far fewer solutions than products within one month. Focusing on the sales generated per month/quarter is therefore rarely appropriate in this case. Instead, the targets that you set your sales employees should be long-term in nature.

Focus of the assessment

Irrespective of the time dimension, the employee assessment can also be carried out on the individual or team level depending on the business model. In the product business, each sales employee has comparable target specifications and is personally responsible for achieving them. Accordingly, assessing performance at the individual level is sensible here.

In the complex solution business, the performance of individual employees is usually no longer comparable. Often, the complex solutions, some of which are high-tech, mean that one individual sales employee can no longer remain familiar with all of the technical characteristics of the offer and explain them to the customer. It is therefore common for the sales teams to join forces and complement one another with regard to their technical knowledge in the solution business. For the sales employee, however, this means that sales success is no longer dependent solely on their own expertise but is also influenced by the performance of the other negotiation participants. Team-level assessment should therefore be given preference over the assessment of individual performances in the solution business.

Case example: What happens in solution selling if sales leadership is not adapted?

If managers maintain their previous management styles when switching to a new business model, it often leads to problems. This is precisely what happened to a company from the mechanical engineering industry. The company had successfully sold production machines for a number of years. Since it was coming under increasing pressure due to digitalization and the increasing number of options for comparing products, however, the management decided to offer services in addition to selling production machines. These included leasing and maintenance contracts as well as predictive maintenance solutions.

As the management had recognized that solution selling requires a different approach than selling products, it had established a new salesforce that was responsible specifically for the sale of solutions. Mr. Miller,[21] a very experienced first-line manager who had previously enjoyed great success in the product business and whom the customers also trusted, was selected as the manager of this new salesforce.

After a few months, the managing directors looked at the sales figures since the launch of the new business model and were disappointed. Although the company was still generating acceptable sales volumes with machines, the sales employees had hardly sold any solution contracts. To find out why, the management arranged a meeting with some of the sales employees. They explained that they had kept to the previously applicable behavioral standards and had attempted to achieve the highest possible sales volume.

The management came to the conclusion that Mr. Miller had continued to manage the new salesforce as he had in the product business. He had set his employees short-term targets and issued clear behavioral specifications. But this management style did not work in the solution business. The sales cycles there were longer and the sales employees had to adapt their behavior more extensively to the respective customers in order to develop individual solutions.

The management then appointed Mr. Smith, a new manager who had experience in solution selling and granted the employees more leeway. The solution salespersons subsequently operated in small teams. If necessary, Mr. Smith got together with his employees following sales pitches in order to determine improvement potentials together. The employees quickly learned which behavior is expedient in the solution business and became more confident in their dealings with solution customers.

[21] The names in this case example have been changed.

WHY SALESTEGY?

STRATEGY – FINDING THE RIGHT BUSINESS MODELS

PRACTICAL EXAMPLES

SALES – IMPLEMENTING NEW BUSINESS MODELS IN SALES

Chapter 12
Adapting lead management

Project steps

- Step 1: Determining customer data
- Step 2: Performing statistical analyses for determining customer potential
- Step 3: Integrating customer potentials in the CRM system
- Step 4: Implementing lead management per business model step

- Deep dive: Implementing social selling in the lead management process

Message of this chapter

Lead management has to be adapted for each new solution business model.

What should you have learned by the end of the chapter?

- You are familiar with the four project steps for adapting lead management for your solution business models.
- You are able to check your lead management for opportunities for social selling approaches.

DEATH VALLEY MARKETS

Strategy of minimum costs and prices

GREEN HILL MARKETS

Strategy of solution business models

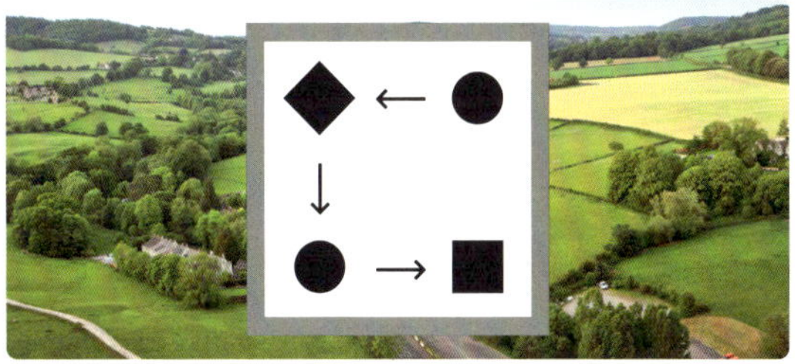

The next area of sales management is concerned with determining customer potentials and the lead management coordinated to this. I deal with four project steps in this process (see Figure 82).

Fig. 82 Implementing new business models – project steps in sales management

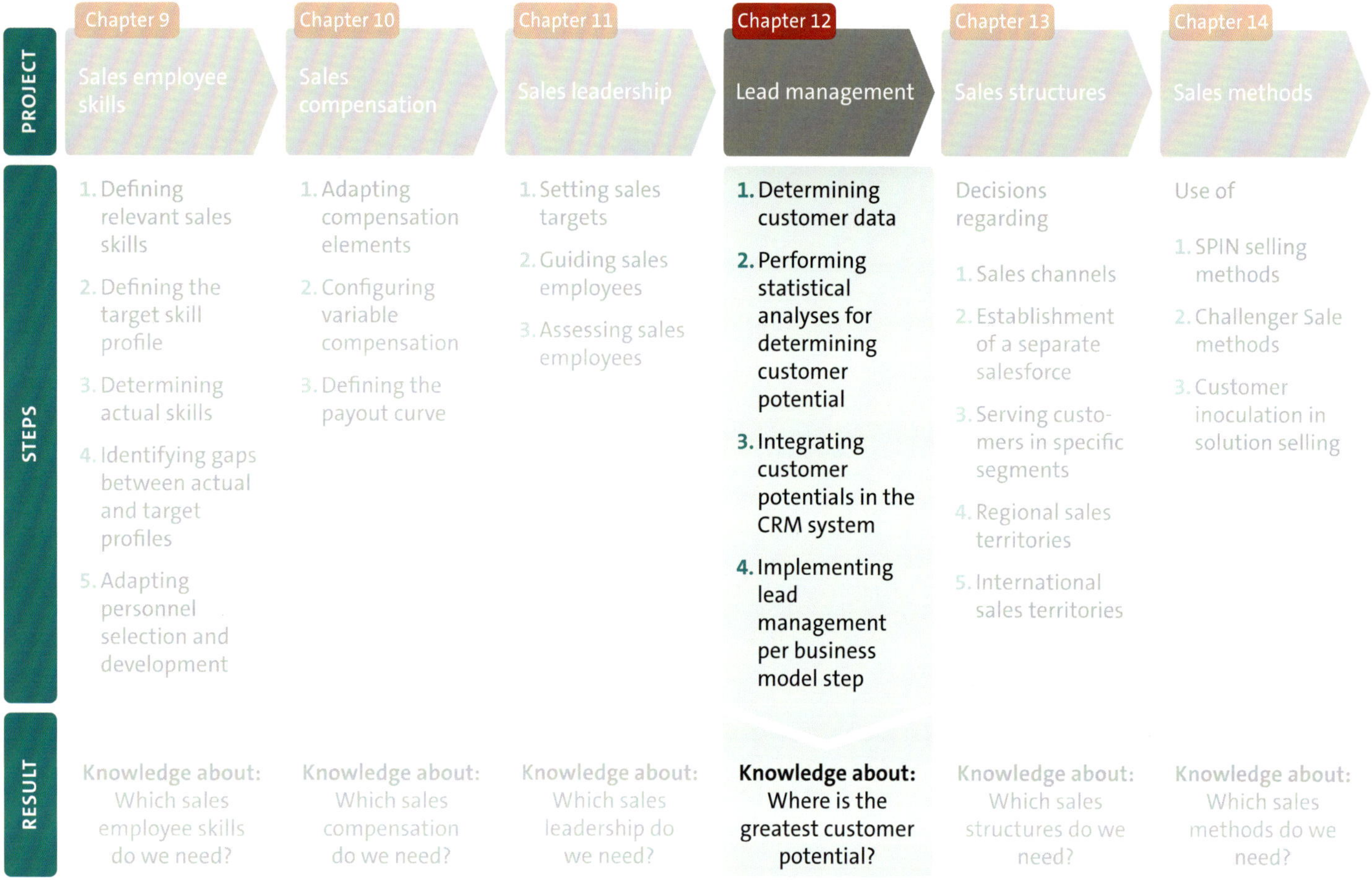

Knowledge of customers' business potential is the basis of lead management. It is therefore vital to first determine how high the level of customer potential is for each new business model. To do this, it is worthwhile applying the logic of the SNOOFY business model audit from Chapter 6 (see Figure 83). There, I described that different business model steps also result in different customer segments that the supplier can serve. The customer potentials first have to be determined on three levels in this case:

1. The overall potential per business model step
2. The potential per customer segment
3. The potential per individual customer

Fig. 83 The three customer potential levels in a business model

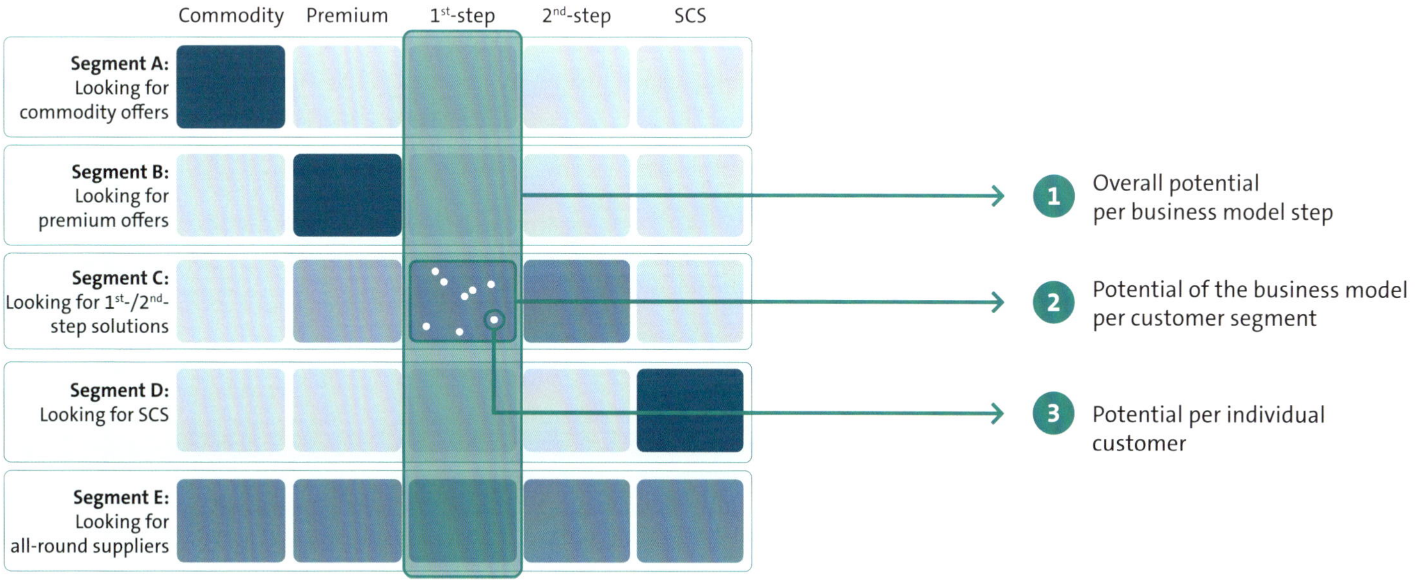

Practice shows that it is neither sensible nor feasible to focus on all segments with the same solution offer. The more precisely the solution potentials per individual customer are known, the more successful selling the solution offers will be. Accordingly, companies that manage to address their offer to precisely those customers with current demand and/or future demand for this offer – and therefore also a corresponding (future) willingness to pay – are successful. One example of adapting the customer potential assessment after entering the solution business is *Normfest* (see information box).

Practical example: Assessment of the solution potential of individual customers

With 400 employees, Velbert-based company *Normfest* sells C-parts and accessories to car workshops. Its customers include numerous small workshops, large car dealerships with several locations and workshops, and industrial companies. In addition to its commodity business, the company has also developed two 1st-step solutions in recent years, namely cost center management and hazardous substance management for car workshops.

However, making these additional services available without differentiating between the entire customer base of more than 30,000 customers would have caused enormous costs and quickly overwhelmed the sales department. Instead, the company attempted to assess the solution potential of individual customers (see Figure 84). Customers with both the demand and a sufficient willingness to pay for the new services were prioritized in this approach.

Fig. 84 Assessment of the customer potential of individual customers

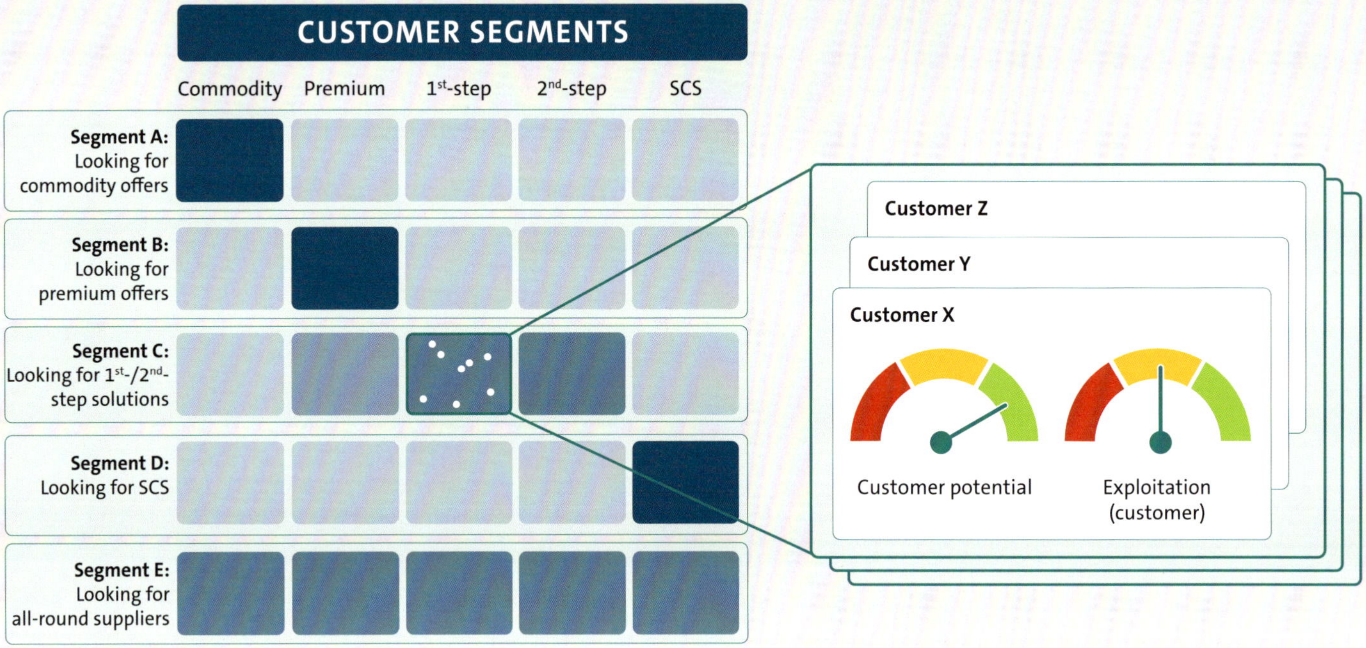

But how do you determine which individual customers are to be handled with priority in solution selling? The classic ABC subdivision of the customers based on their previous sales volume is a widely used approach here. This subdivision then results in conclusions for sales management as regards the frequency of visits or the pricing, for instance.

However, this procedure quickly encounters its limits when switching to solution business models. How are we to draw conclusions regarding a customer's *total* potential with commodity products, or even with 1st-step solutions, based on its current sales volume in the commodity sector, for instance? Ultimately, the customer's current sales volume tells us nothing about either its total demand (the customer may obtain the majority of its services from the competition) or its future demands, which may possibly arise due to market developments or investment decisions. If sales were to be managed solely according to customers' current actual sales volumes, customers that might currently be smaller but may potentially be very valuable would be completely disregarded.

Our consulting work has resulted in a four-step scheme for adapting lead management: 1) Determining customer data, 2) performing statistical analyses for determining customer potential, 3) integrating customer potentials in the CRM system, and 4) implementing lead management per business model step. I describe the individual steps in detail further on in this section.

Project steps

Step 1: Determining customer data

The challenge in the first step of analyzing customer potential lies in obtaining and bringing together a variety of data that can be assumed to be related to the customer potential. A distinction has to be made between two types of data in this process. More qualitative data – or data that cannot be easily expressed in numbers – is also important, in addition to easily quantifiable data such as sales volume, number of employees, balance sheet total, and investment volumes. The former includes assessments of the customer's digital maturity, its purchasing professionalism, and the sustainability of its business model. This makes obtaining the data difficult.

In addition, many companies use different platforms, systems, and department- or country-specific processes, which often makes determining this information very laborious. However, the good news is that the salesforce is an outstanding source of both quantitative and qualitative data. This means that a field sales representative who is familiar with a customer is aware of both the hard facts about this customer company and characteristics that are to be evaluated more subjectively. However, a fundamental problem arises from a sales management perspective: This knowledge is only available in decentralized form in the majority of cases, and it can only rarely be seen by the management. It is therefore necessary to recover this knowledge and, after corresponding preselection, to ask the sales employees to supply customer data that provides information about the customer potential.

This data then has to be collected based on a random sample of customers. The random sample should encompass around 10% of the actual customer base, but at least 100 customers that are as representative as possible. An example of this is provided in the information box.

Practical example: The road to determining customer potentials

With around 16,500 employees and a sales volume of approximately $2.6 billion, *Phoenix Contact* supplies a number of different products in the electrical engineering and automation sector. At present, *Phoenix Contact* is also in the midst of expanding from the extensively product-centric business into selling 1st- and 2nd-step solutions. These include solutions that enable machine remote maintenance, for instance. From the physical components to the appropriate transmission paths and software solutions, *Phoenix Contact* is able to offer everything from a single source. However, by no means does every customer also have the corresponding demand potential for such solutions. It is therefore extremely important for the company to deploy its sales resources as well as possible according to customer potentials.

The obstacle in this case also involves systematically determining customer potentials. Optimally, it should be possible to express these as the most accurate possible estimate of a customer's sales volume potential (in dollars) and should give consideration to all of *Phoenix Contact's* business model steps. As the customers are very heterogeneous and also purchase from completely different *Phoenix Contact* product segments, an entire range of approaches for determining potential had already evolved historically, so to speak. To do this, for instance, external industry data or invitations to tender were analyzed or the sales employees were surveyed to be able to predict potentials more or less precisely. However, these fragmented approaches were problematic for the sales management because the resulting findings could not be either transferred to all sectors or used for effective resource management.

This therefore led to a joint project with *Phoenix Contact* Germany that was headed by my colleague Professor Christian Schmitz. In this project, we used a uniform procedure across all business model steps to develop a systematic potential analysis. This involved enabling an automated potential analysis in the CRM system that allows the potential of the individual customer to be estimated (in dollars) without a great deal of effort based on the customer data available in the CRM system. In this project, additional one-off efforts were needed to collect the data required in the first step using external data sources. Almost 100 different potential indicators were identified, evaluated, and prioritized. Ultimately, around 20 of these indicators were collected and included in the statistical analyses that are described in step 2.

Step 2: Performing statistical analyses for determining customer potential

The important work comes after the data has been collected. Various statistical approaches (such as regression and time series analyses) can now be used to determine which information criterion provides significant explanatory power for the customer potential. Data prioritization can then be carried out according to the level of explanatory power. The lower area of Figure 85 shows example figures that illustrate the possible results of such an analysis.

Fig. 85 Statistical analyses for determining customer potential

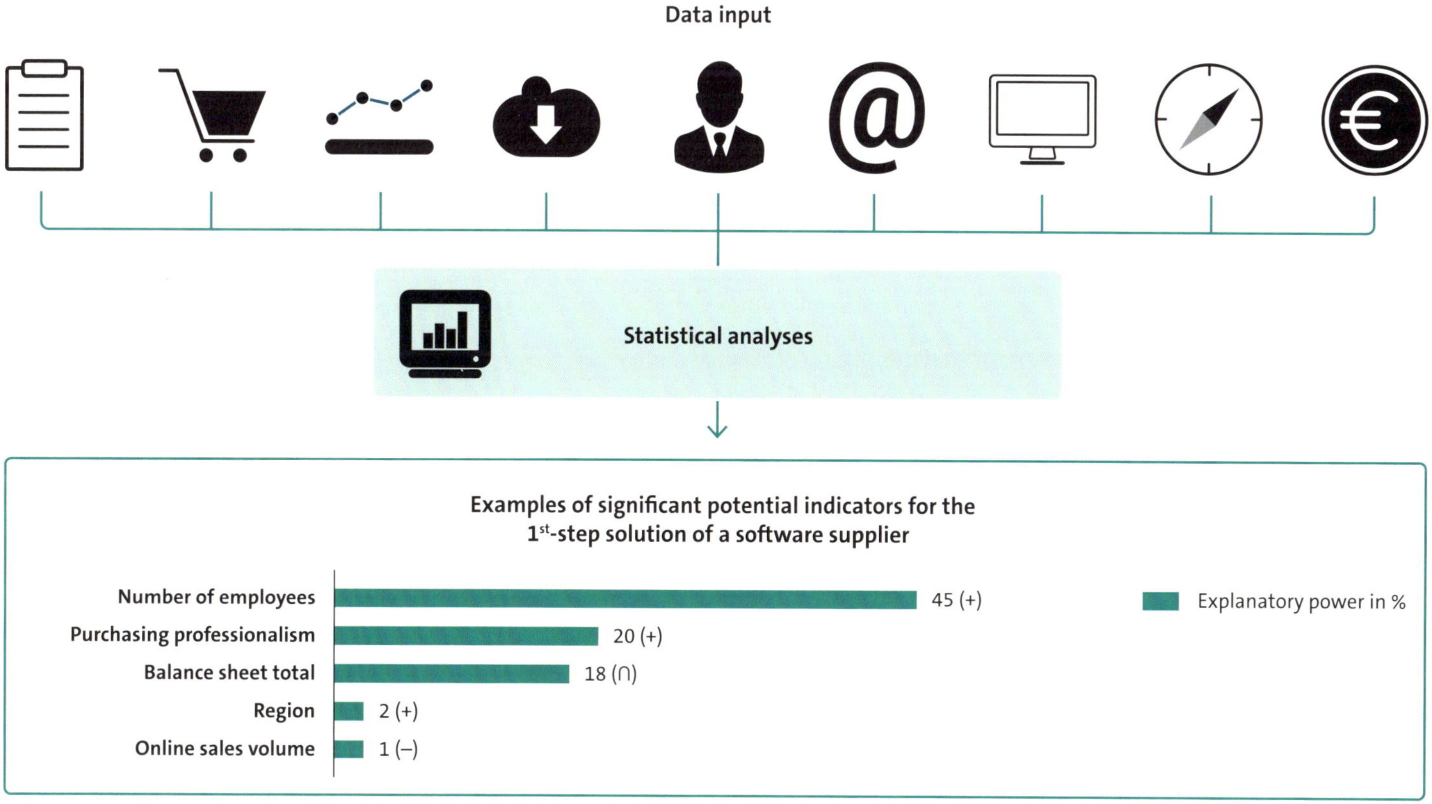

In our analyses, we have often seen that the customer's number of employees is a significant indicator of its potential. In the example of the software supplier in Figure 85, it is 45%. It is therefore little wonder that many companies measure customer potentials based on the number of employees. This functions particularly well when the customers operate very similarly within a segment with reference to their business models.

However, that is not always the case. As an example, let us look at a supplier of software for stockkeeping that is focused on construction trade companies. If it estimates customer potentials based exclusively on the number of employees, it follows the assumption that companies with more employees also require more software licenses, as they may possibly have several locations and have to manage larger warehouses. So far, so good. However, the fact that there may be major differences in the degree of professionalization within the customers' businesses is disregarded. Some customers may use their own Excel-based solutions or have outsourced warehouse management.

It is therefore advisable for the supplier to include further potential indicators in its customer analysis. For instance, the professionalism of the customer's purchasing department could be evaluated by sales. In our example, this information provides an important indication of customer potential – still with explanatory power of 20% (see Figure 85).

In our example, the role of the balance sheet total is interesting. An inverse U-shaped relationship was determined here. Figure 86 shows this relationship: Up to a certain point, the customer potential grows as the balance sheet total increases. If this point is exceeded, the potential decreases again. This could be explained by the fact that, in the case of very high balance sheet totals, it is more likely for a company to provide certain services in-house rather than buying them.

It may therefore be important as well to check the data for such relationships that are more complex. It is technically simple. If they are taken into consideration, the sales hit rate increases later on.

Fig. 86 The relationship between the balance sheet total and customer potential (example of a software supplier)

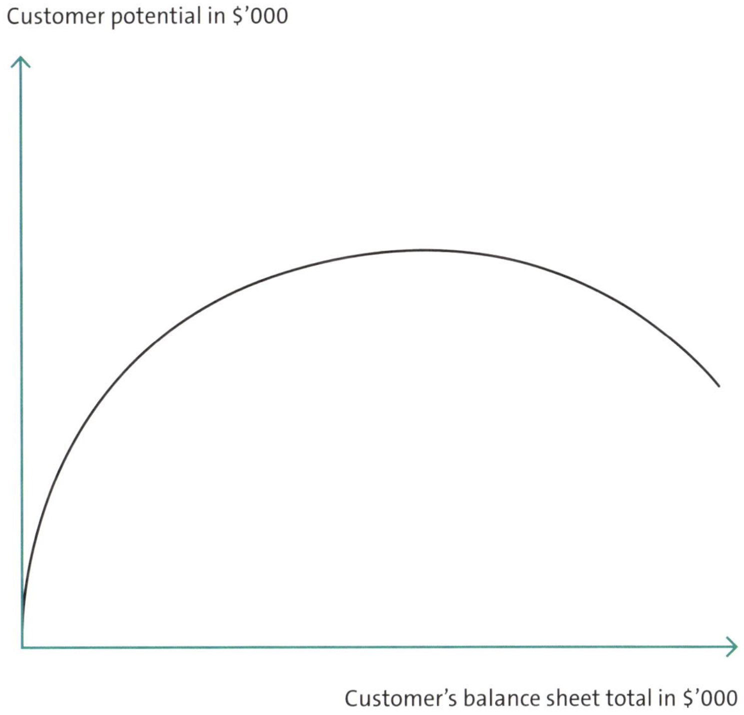

Once the statistical relationships in the data have been determined, they have to be transferred to a potential analysis model. Such a model corresponds to a mathematical algorithm into which all of the factors for calculating the customer potential are input with an optimal weighting. In order to adjust the weighting, the relationships are checked based on a second (usually slightly smaller) random sample of customers. After all, it may be the case that the statistical analysis has only illustrated the characteristics of a specific group of customers. The check can be used to further refine the potential analysis model and finally implement it.

In order to then use it universally, the supplier must collect the relevant data for the entire customer base and calculate the customer potential for each customer. That sounds time-consuming, but we have often done it together with companies.

In summary, it can be said that this procedure can be used to predict the customer potentials of existing products/solutions in field sales. The potential model can also be used to quickly qualify and prioritize new customers using fewer criteria.

Step 3: Integrating customer potentials in the CRM system

A benefit can only arise if the salesforce works with the potential information on a daily basis, as the sales employees then invest more time in high-potential customers. To make the information accessible for daily use, the potential analysis model must be implemented in the CRM system.

Today, it is possible to implement it so that the potential analysis continually updates itself and becomes increasingly precise thanks to new data points.

Figure 87 shows an example of a potential analysis tool. This tool enables a quick overview of the potential of a specific customer. The sales employees should take this potential into consideration when planning their journeys and preparing for visits. At the same time, the tool shows the exploitation of the potential thus far. In our example, the potential has only been exploited to a slight extent. This therefore involves a customer that can be further developed. Each of these analyses can be shown in greater detail using degrees of potential exploitation per month (at the bottom of the figure), for example.

Fig. 87 Example of a customer potential analysis tool for sales employees

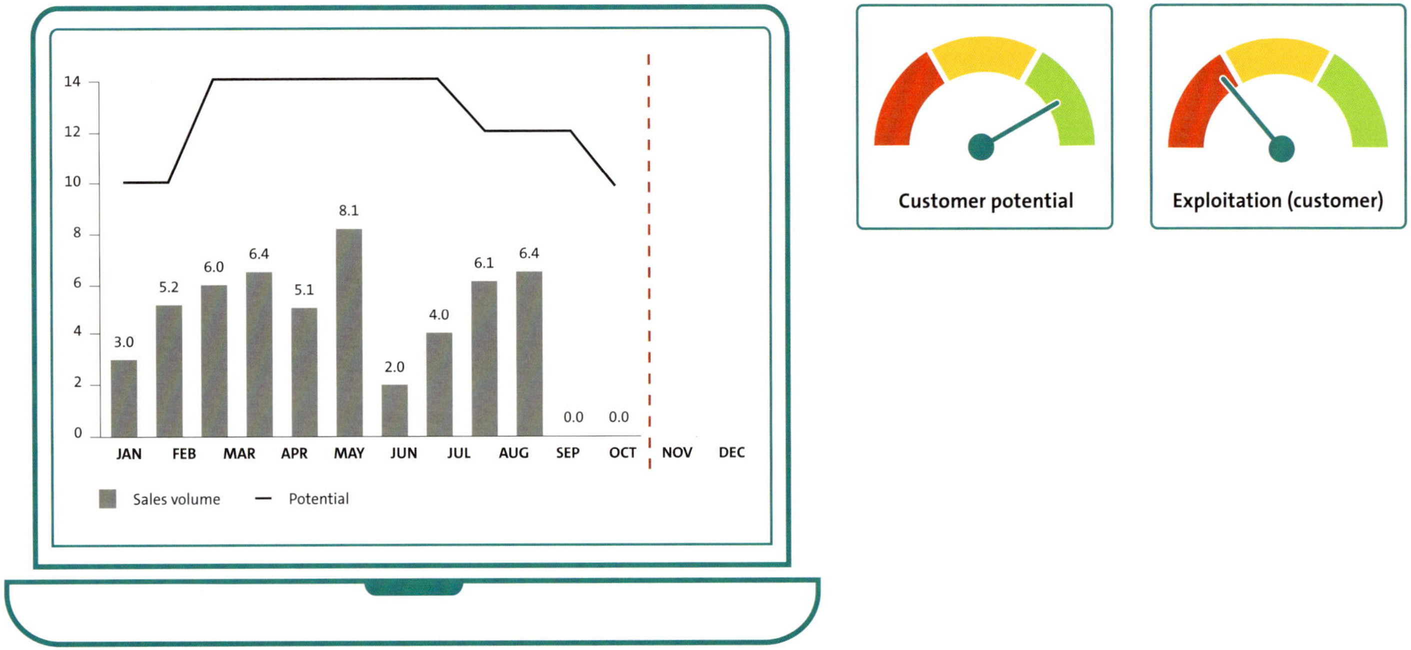

Information box: Startups can also determine customer potentials

Companies face a very particular challenge whenever they want to ascend to the next business model evolutionary step: They have to launch a new offering with as of yet unknown sales volume potential onto the market and, in some cases, do not even know which customers are willing to pay for the solution. This situation is entirely comparable with that of startups. They, too, are often unable to make use of empirical values.

This can be clearly illustrated using the example of the startup *PHYSEC*. *PHYSEC* develops innovative solutions for connecting devices of all kinds to the Internet (key term "Internet of Things"). Among other things, the company has developed a chip that can be both retrofitted in existing devices and also installed in new devices during production. The application area of this technology ranges from very simple goods (such as coffee makers) up to and including highly complex systems (such as cooling systems in nuclear power plants). The chip is then able to transmit data to the production company via a high-security cryptographic link, and thus enable services such as predictive maintenance as well as supplying information about customers' usage behavior, which is important for product development. Such chips also enable *outcome-based pricing*, as the chip is able to register performance data and transfer it in real time. *PHYSEC's* solution therefore necessitates a certain level of technological maturity on the part of the customer company.

As exciting as the business model idea is, it was initially very difficult for the startup to identify possible customers. If existing demand had been used as the sole yardstick, the market potential would have been comparatively low. At the time when *PHYSEC* was founded, only a very few companies had digitalized their products. However, the future potential was enormous, so the sales strategy involved consciously contacting companies that were not yet using such chips but were essentially open to this technological innovation.

But how could such companies be identified? To precisely analyze the potential, *PHYSEC* first drew up a list of criteria that qualify a customer in terms of the new services. In this case, creativity regarding possible application areas was required. The first consideration was that the chips should basically be of interest to all companies that manufacture physical devices with electronic components, as they can be installed there and used to collect data. As the issue of high-security data transfer, in particular, was *PHYSEC's* special feature, the company struck manufacturers of comparatively low-value products from the list of potential customers and focused more intensively on premium manufacturers. Companies with a great affinity for research and development that occupied pioneering roles in their respective industries were additionally sought. *PHYSEC* used these criteria to specifically draw up lists of potential customers and then prioritize and qualify them again based on the established criteria. Finally, *PHYSEC* attempted to specifically make contact at trade fairs with those customers in which they had identified a particularly high level of potential. This approach was very successful. *PHYSEC* not only won numerous startup competitions and millions of dollars in subsidies, but it is now also working successfully together with numerous high-potential customers. Incidentally, the process of drawing up such potential lists can now be carried out almost completely automatically thanks to search and evaluation algorithms.

Step 4: Implementing lead management per business model step

In the fourth step, the supplier integrates the information about the customer potential into a lead management process. This process should be thought through separately for each business model step.

The lead management process describes the selection process used to determine a significantly smaller number of *leads* from an initially high number of *suspects*. The supplier must also analyze this process separately for each business model step. Figure 88 shows such a lead management process along the sales funnel steps *(contact, suspect, prospect, lead, customer)*.[22]

However, this does not mean that all of the funnel steps always have to be run through. In many cases, individual steps are skipped. By being addressed pro-actively, for example, a suspect can also directly become a *lead*. In addition, the sales funnel processes are often iterative. This means that if *leads* do not immediately become *customers*, they can be thrown back "into the pot" as *prospects*.

Contact: Establishing a pool of contacts

The first objective in the lead management process is to use various channels to generate a large pool of unqualified contacts that could potentially become customers. As can be seen in Figure 88, companies undertake research of varying breadth or focus to do this, depending on the business model step.

- **Commodity and premium products:** Due to their low level of complexity, a high number of potential customers usually arises for commodity and premium products. This results in very broad-based contact research.
- **Solutions:** By design, solution selling is only relevant to a smaller group of customers, since sales are dependent on integrating the solution into the customer's value creation. Therefore, 1st-step solutions already necessitate focused research, which becomes increasingly important in the case of 2nd-step solutions and super-complex solutions.

Suspect: Classification on the basis of customer potential

The search processes result in an unqualified pool of possible customers that are generally referred to as *suspects* in the sales funnel. In order to determine a focused group of promising potential customers, the contacts are classified on the basis of their estimated customer potential in this step.

[22] I use one of the common nomenclatures of the sales funnel steps here. In practice, we often experience discussions concerning the "right" names of the steps. They arise because the designations are dependent on both the history of the company and the CRM system used in each case. However, a simple subdivision into the above five steps is sufficient for the following description of lead management per business model step.

Fig. 88 Lead management per business model step

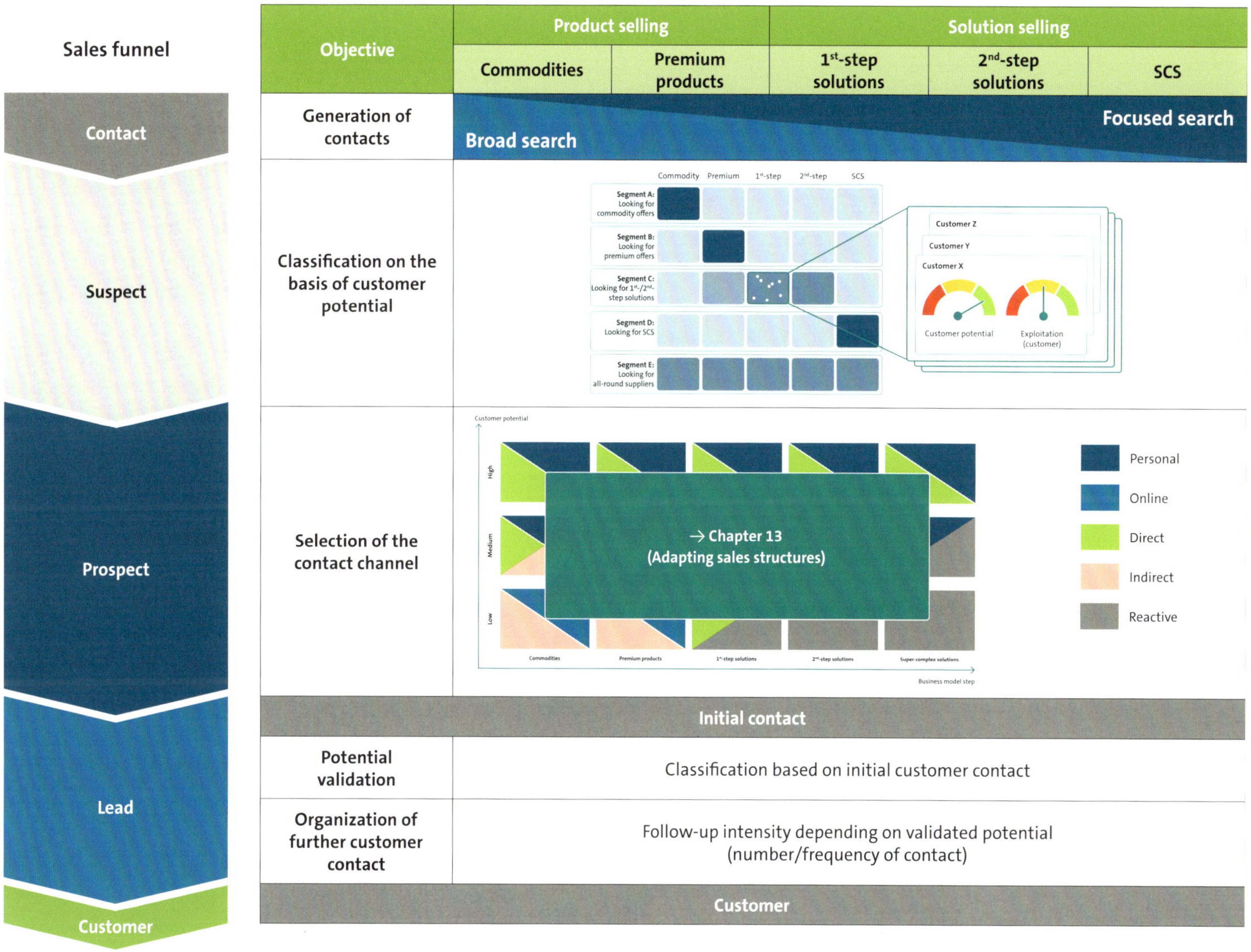

Prospects: Selection of the contact channel

The pool of promising contacts remaining on the basis of customer potential is termed *prospects*. Based on the estimated customer potential, the supplier now selects the appropriate channel for initial contact. In addition to the customer potential, the choice of channel depends on the business model step, each of which has different requirements (see Figure 89).

Fig. 89 Sales channel structure per business model step

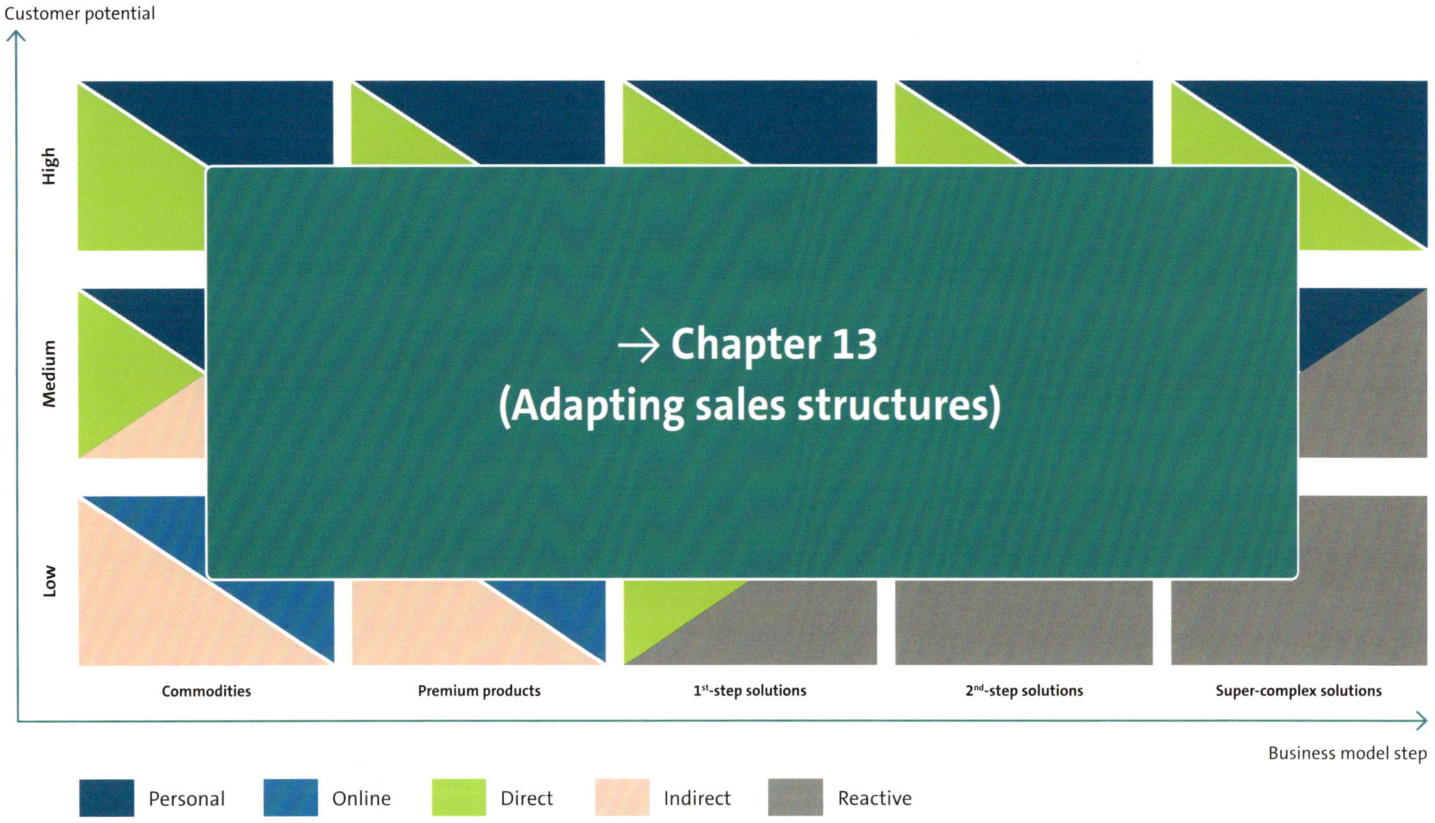

Leads: Potential validation and organization of further customer contact

Following initial contact, the supplier compares the customer's reaction and the indicated purchasing interest with the initially assumed customer potential. Depending on the result, it will define the customer as a *lead*. The follow-up intensity and whether a separate sales employee is used for the customer depend on the validated potential in each case. This results in an individual approach for each customer – depending on the business model step – and includes the frequency of customer contact and acquisition expenses.

Deep dive: Implementing social selling in the lead management process

Social media is becoming increasingly important on both the supplier and customer sides in B2B business. Accordingly, sales organizations are increasingly using social selling in the lead management process.[23] This primarily involves addressing decision-makers from the customer company. It must be taken into consideration that several decision-makers are often involved in purchasing many B2B business environments.

Social selling can essentially take place *passively* (without interaction with other social media users) or *actively* (with such interaction).

- *Passive:* Passive social selling refers to the collection of publicly available information (such as profile information, forum contributions), which is then used in selling. Numerous application aids are available to sales employees for this (the LinkedIn Sales Navigator, for example).
- *Active:* Sales employees can also undertake active social selling by writing their own contributions, notifying other users, networking with other users, and individually exchanging information with other users.

Figure 90 shows how social selling and the sales funnel shown above *(suspects, prospects, leads*, and *customers)* can be combined. In *passive social selling*, the supplier focuses on determining the following information about the customer: buying signals, purchasing-related needs, and personality traits. To do this, the sales employees can evaluate the customer decision-makers' personal profiles. This information helps the supplier to register customer potential, classify it into the sales funnel categories, and prepare for contact.

In the next step, the social selling activities have to be applied to the systematology of the business model steps. Figure 91 shows the weighting of individual social selling activities using the example of a telecommunications group.

[23] Focus is placed on social selling in this deep dive. Of course, there are also other highly relevant lead channels such as "call-to-action" content on websites (white papers, videos, etc.), expert presentations at industry association events, acquisition at specialist trade fairs, referral marketing via existing customers.

Fig. 90 Social selling activities along the sales funnel

Passive social selling

Sales funnel \ Information type	**Buying signals** (e.g., specific purchasing question as a contribution in a LinkedIn group)	**Purchasing-related needs** (e.g., exchange of experience about product characteristics in the comments of a YouTube video)	**Personality traits** (e.g., degree of security orientation based on indications in the personal profile)
Suspects	**Determination through: Filling the sales funnel with a variety of potential customers**		
Relevance of the information type	★☆☆☆☆	★★☆☆☆	☆☆☆☆☆
Social selling activity	**Activity?** Broad-based research **Reference point?** Variety of company profiles, personal profiles, and contributions **Research content?** Initial indications regarding buying signals and purchasing-related needs (e.g., content of contributions/discussions about products)		
Prospects	**Determination through: Rough screening according to indications regarding purchasing-related needs and buying signals**		
Relevance of the information type	★★★☆☆	★★★★★	★☆☆☆☆
Social selling activity	**Activity?** Focused research **Reference point?** Preselection of personal profiles and contributions **Research content?** Search for indications regarding purchasing-related needs and buying signals as well as initial inspection of the personal profile (e.g., "I am looking for" data in the LinkedIn personal profile)		
Leads	**Determination through: Fine screening based on more detailed information on personality, purchasing-related needs, and buying signals; preparation for making personal contact with leads**		
Relevance of the information type	★★★★★	★★★★★	★★★☆☆
Social selling activity	**Activity?** Highly focused research **Reference point?** Specific selection of personal profiles and contributions **Content?** Specific personal information and occasions (e.g., very pronounced buying signal)		
Customers	**Determination through: Successful conclusion of a purchase; then: Granular screening using changing or supplementary information (updating)**		
Relevance of the information type	★★★★☆	★★★★☆	★★☆☆☆
Social selling activity	**Activity?** Repeated research **Reference point?** Comparatively low number of personal profiles and contributions **Research content?** Changing or complementary buying signals, purchasing-related needs, and personality traits		
Applications for collecting this information (examples)	• Reddit Keyword Monitor Pro • Geopiq for Instagram • Hootsuite Insights	• Echobot • TweetDeck	• LinkedIn Sales Navigator • etc.

Occasions for making contact (e.g., profile news, anniversary, current studies) as the link between passive and active social selling

Feedback of information to adapt passive social selling

Active social selling (examples)

Sales funnel	Active social selling (examples)
Suspects	• Responses to public forum contributions Objective: Additional information collection
Prospects	• Responses to public forum contributions Objective: Additional information collection • Creation of forum contributions Objective: Generation of attention and demonstration of expertise
Leads	• Personal communication within social media Objective: Information exchange, additional information collection and relationship establishment • Networking with decision-makers Objective: Additional information collection and relationship establishment
Customers	• Personal communication within social media Objective: Information exchange, additional information collection and relationship management • Recommendations from or for decision-makers Objective: Network expansion

Fig. 91 The weighting of social selling activities along the business model steps using the example of a telecommunications group

Activity areas		Commodities	Premium	1st-step solutions	2nd-step solutions	Super-complex solutions	Basis for tendency classification
Passive social selling	**Objective**	Importance of passive social selling					**Consistent importance of passive social selling**
1 **Collection of information** that is publicly available	Screening out irrelevant prospects	★★★★☆	★★★★☆	★★★★☆	★★★★☆	★★★★☆	• Information collection is equally relevant for all business models • It enables prospective potential analyses • It enables leads to be addressed individually
	Preparing initial contact	★★★★☆	★★★★☆	★★★★☆	★★★★☆	★★★★☆	
Active social selling	**Objective**	Importance of active social selling					**Increasing importance of active social selling**
2 **Write own contributions** (i.e., write new contributions or respond to contributions)	Generating attention	★★☆☆☆	★★☆☆☆	★★☆☆☆	★★☆☆☆	★★☆☆☆	• To a certain extent, generating attention is relevant in all business models in order to be noticed by the generally high number of decision-makers in the first place • However, expertise is comparatively more important to attract decision-makers, particularly in the case of complex solutions
	Establishing and managing relationships						
	Demonstrating expertise	★★★☆☆	★★★☆☆	★★★☆☆	★★★★☆	★★★★☆	
3 **Notifying other users**	Generating attention	★★★☆☆	★★★☆☆	★★☆☆☆	★★☆☆☆	★★☆☆☆	• In the product sector, standardized messages are a simple way of attracting the attention of decision-makers, but they can also cause reactance • Individualized messages are primarily necessary in the solution sector to establish relationships with specific decision-makers and demonstrate expertise to them
	Establishing and managing relationships	★☆☆☆☆	★☆☆☆☆	★★☆☆☆	★★★☆☆	★★★★☆	
	Demonstrating expertise	★☆☆☆☆	★☆☆☆☆	★★☆☆☆	★★★☆☆	★★★★☆	
4 **Networking with** other **users**	Generating attention	★★☆☆☆	★★☆☆☆	★★★☆☆	★★★☆☆	★★★☆☆	• Contact requests can be an option for attracting the attention of decision-makers and require hardly any work • However, they are more important for the relationship establishment, which is particularly significant in increasingly complex business models
	Establishing and managing relationships	★★☆☆☆	★★☆☆☆	★★★☆☆	★★★★☆	★★★★☆	
	Demonstrating expertise						
5 **Individually exchanging information** with other users	Generating attention						• In increasingly complex business models, individual exchange is crucial for establishing relationships and demonstra-ting expertise • However, personal exchange in the product sector is usually very laborious due to the generally high number of customers
	Establishing and managing relationships	★★☆☆☆	★★★☆☆	★★★★☆	★★★★★	★★★★★	
	Demonstrating expertise	★☆☆☆☆	★☆☆☆☆	★★★☆☆	★★★★☆	★★★★☆	

SALES – IMPLEMENTING NEW BUSINESS MODELS IN SALES

Chapter 13
Adapting sales structures

Project steps

- Step 1: Decisions regarding the sales channels
- Step 2: Decisions regarding the establishment of a separate salesforce
- Step 3: Decisions regarding serving customers in specific segments

- Deep dive: Is personal selling always fruitless in commodity markets?

- Step 4: Decisions regarding regional sales territories
- Step 5: Decisions regarding international sales territories

Message of this chapter

The sales structures have to be adapted for each new solution business model.

What should you have learned by the end of the chapter?

- You are familiar with the five project steps with which you can adapt the sales structures for your solution business models.
- You are able to check whether – and which – potentials have remained for personal selling in the commodity markets.

DEATH VALLEY MARKETS

Strategy of minimum costs and prices

GREEN HILL MARKETS

Strategy of solution business models

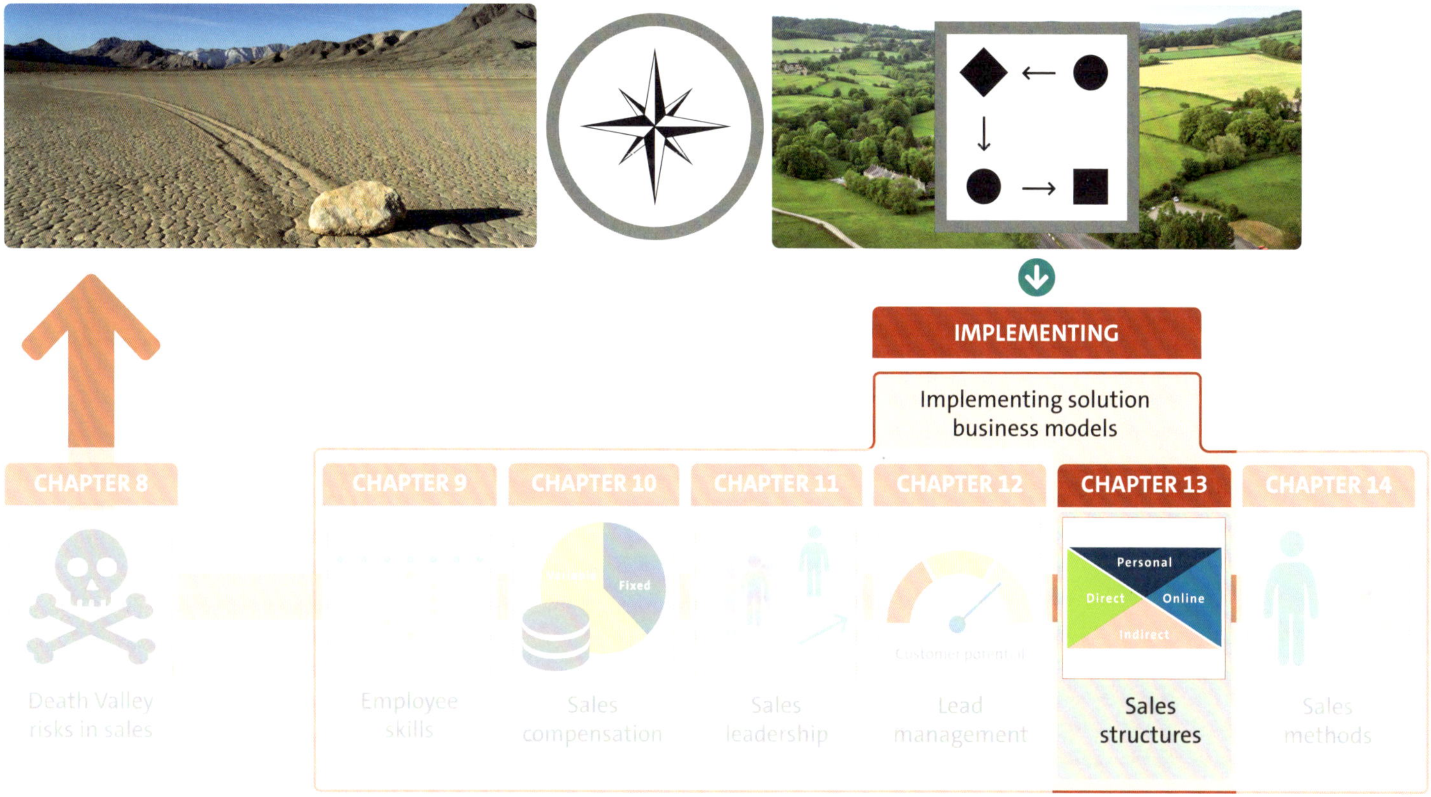

The sales structures are another sales management area that often causes solution business models to fail. In our project steps, the supplier has to make five decisions (see Figure 92):

Fig. 92 Implementing new business models – project steps in sales management

	Chapter 9	Chapter 10	Chapter 11	Chapter 12	Chapter 13	Chapter 14
PROJECT	Sales employee skills	Sales compensation	Sales leadership	Lead management	Sales structures	Sales methods
STEPS	1. Defining relevant sales skills 2. Defining the target skill profile 3. Determining actual skills 4. Identifying gaps between actual and target profiles 5. Adapting personnel selection and development	1. Adapting compensation elements 2. Configuring variable compensation 3. Defining the payout curve	1. Setting sales targets 2. Guiding sales employees 3. Assessing sales employees	1. Determining customer data 2. Performing statistical analyses for determining customer potential 3. Integrating customer potentials in the CRM system 4. Implementing lead management per business model step	Decisions regarding 1. Sales channels 2. Establishment of a separate salesforce 3. Serving customers in specific segments 4. Regional sales territories 5. International sales territories	Use of 1. SPIN selling methods 2. Challenger Sale methods 3. Customer inoculation in solution selling
RESULT	Knowledge about: Which sales employee skills do we need?	Knowledge about: Which sales compensation do we need?	Knowledge about: Which sales leadership do we need?	Knowledge about: Where is the greatest customer potential?	Knowledge about: Which sales structures do we need?	Knowledge about: Which sales methods do we need?

The following table enables you to conduct a quick test on your sales structures. You can run through the specified steps for each new business model and make decisions regarding changes to your sales structure.

Quick test on sales structures for a new business model

Step	Question	Yes	No
Step 1: Decisions regarding the sales channels 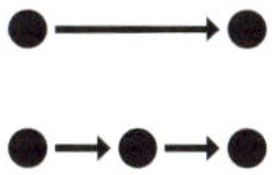	**Question 1:** Does the customer benefit in the relevant business model require extensive explanation?	☐	☐
	Question 2: Is there a high level of uncertainty in the relevant business model from the customers' point of view?	☐	☐
	→ 2 × *no*: Indirect/online selling possible		
	→ At least 1 × *yes*: Direct selling with own salesforce necessary		
Step 2: Decisions regarding the establishment of a separate salesforce 	**Question 1:** Are the skills necessary for selling in the envisaged business model ***available*** in the existing salesforce?	☐	☐
	Question 2: Can the skills necessary for selling in the envisaged business model ***be learned*** by the existing salesforce?	☐	☐
	→ Question 1 *no* and question 2 *yes*: Implement training courses/coaching		
	→ 2 × *no*: Establish new salesforce, implement team selling		
Step 3: Decisions regarding serving customers in specific segments 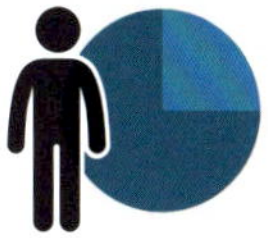	**Question 1:** Is the business volume large enough to process individual customer segments with separate sales units?	☐	☐
	Question 2: Is the necessary segment-specific know-how available in your salesforce?	☐	☐
	→ Question 1 *no*: Do not segment customers		
	→ Question 1 *yes* and question 2 *no*: Further training to develop sales employees into segment experts or recruitment of sales employees with industry know-how		
	→ 2 × *yes*: Subdivide salesforce according to customer segments		

Step	Question	Yes	No
Step 4: Decisions regarding regional sales territories 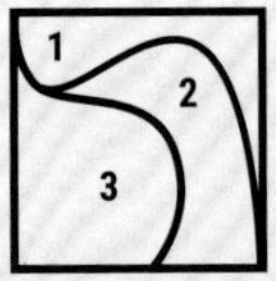	**Question 1:** Is the customer potential in the territories comparable?	☐	☐
	Question 2: Is the territory breakdown optimized in terms of costs?	☐	☐
	→ At least 1 × *no*: Revise territory breakdown		
	→ 2 × *yes*: Retain territory breakdown		
Step 5: Decisions regarding international sales territories 	**Question 1:** Is the market potential for the business model adequate in the target country?	☐	☐
	Question 2: Are the barriers to entering the target country surmountable?	☐	☐
	Question 3: Does your salesforce have the necessary skills to sell the business model successfully on the ground?	☐	☐
	→ Question 1 or 2 *no*: No entry into the national market		
	→ Questions 1+2 *yes* and question 3 *no*: Cooperation with a sales company in the location		
	→ 3 × *yes*: Entry into the national market with own salesforce		

Project steps

Step 1: Decisions regarding the sales channels

As regards the sales structure, you first have to decide which main sales channel is to be used. This is therefore concerned with whether selling is to take place online or offline and whether a direct or indirect approach is to be chosen.

It must be stated in advance that customers on all business model steps are increasingly coming to expect the omni-channel availability of suppliers. They essentially assume that they can reach the supplier at all times via various routes: Online, via social media, email, call center, or in person. Omni-channel management has become a basic element of serving customers.

Sales channels: Online vs. personal selling. In terms of the main sales channel, however, we can see a clear split. The online channel will become even more relevant for product-centric business model steps (commodity and premium products). Conversely, personal selling is still recommended as the primary sales channel for the three solution steps (1st-step, 2nd-step, and super-complex solutions).

This is an average assessment of all of the companies that we studied. It is essential to individually determine the optimal sales channel structure for each company, each business unit, each business model, and each customer segment (see also the deep dive from p. 317).

But *why* is the online channel more suitable as the main channel for one business model and personal selling for others? We have analyzed several parameters concerning this in our studies. In the process, we have come across two elements that have already been described as barriers to entering solution business models in Chapter 2: the necessity of explaining the benefit to customers and customer uncertainty or the purchasing risk that customers perceive. Both elements are pivotal to selecting the sales channel. Our quick test for sales structures (p. 305) therefore includes the following two questions:

Question 1: Does the customer benefit in the relevant business model require extensive explanation?

Question 2: Is there a high level of uncertainty in the relevant business model from the customers' point of view?

If only one of the questions is answered *yes*, direct selling with a separate salesforce is sensible.

There are several findings for both of the questions in our five business model steps. As can be seen in the upper area of Figure 93, they can be arranged between two extreme poles based on the necessity of explaining the benefit and the customer uncertainty. The farther a business model moves to the right, the more its benefit requires explaining and the higher the perceived purchasing risk becomes. The business model step arrangement in the figure shows just how large the distance between the product-centric business models (commodity and premium) and the solution business models (1st-step, 2nd-step, and super-complex solutions) is.

Fig. 93 The necessity of explaining the benefit and the customer uncertainty in the five business model steps

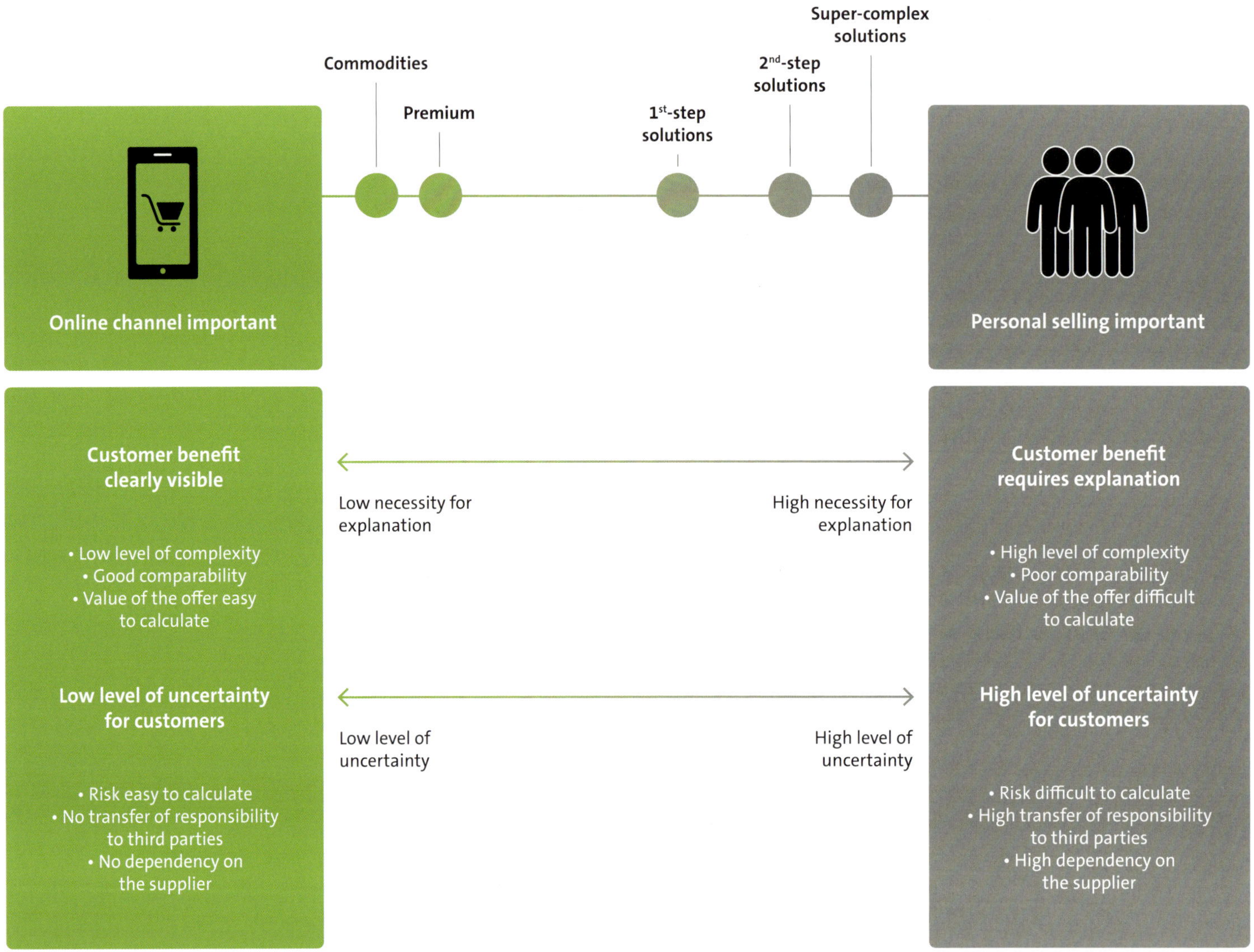

In each case, three characteristics determine whether the necessity of explaining the benefit and the levels of customer uncertainty are high or low. In the case of commodities, the *customer benefit is clearly visible*, as these products are less complex and easily comparable. Here, the value of the offer is often derived from the production costs. Conversely, solution offers derive their benefit due to their high degree of individualization and the high level of service content (see **S**NOOFY – ***S****ervitization*). This inevitably leads to a level of complexity in which comparing the offers with competing offers is barely possible. As a result of this, the entire decision-making process becomes more difficult for the customer to structure, and the value of the offer is harder to calculate.

The *customer uncertainty* is also determined by three characteristics. Let us look at the solution business models to analyze this. In these models, the risk of quality defects is barely calculable compared to the product sector. As described above, the time of the purchase decision plays an important role here (see SNOOF**Y** – ***Y****ield customer value-in-use*, p. 121). In the classic product business, an initial quality analysis can take place prior to the purchase since the product already exists. In the solution business, however, the purchase decision has to be made prior to service provision, and the buyer does not know how good the result will be. Added to this is the fact that responsibility is increasingly transferred to the supplier during the evolution from 1st-step to super-complex solutions (see SN**O**OFY – ***O****perate processes for customers*). Consequently, the customer is extensively dependent on the supplier, which in turn leads to increased customer uncertainty.

Direct vs. indirect selling. The question of whether a direct or an indirect sales approach is to be used is also raised. Do you undertake all sales activities with your own company's resources, or do you outsource selling in whole or in part to external channel partners such as wholesalers and specialist retailers, commercial agents, and online platforms?

The answer to this question is simple. Indirect sales channels are often sensible in the product business. Wholesalers and other channel partners play an important role here. Conversely, indirect selling is not very appropriate for the solution business models. The requirements of the salesforce are simply too extensive.

The service franchise networks found, inter alia, at software suppliers can be an exception to this rule. However, franchise networks are found more seldom as the importance of the SNOOFY factors ***O****perate processes for customers* and ***F****ar-reaching relationship* increases.

Direct access to the customer is necessary in order to steer the salesforce in the right direction for these complex business models and to retain the valuable process know-how. Outsiders cannot usually take over the complex management and development of the salesforce in an equivalent manner. This is true for both super-complex solutions and also for 1st- and 2nd-step solutions.

The question of direct or indirect selling is also raised for the online sales channel. Should selling be carried out exclusively via one's own online channel (directly) and/or **indirectly** via other online dealers and platforms? The latter is already universally available in the B2C sector, whereas the online trading platforms are just gaining ground in the B2B sector.

Let us combine our findings thus far. In Chapter 12, we dealt with the customer potential as the basis for the way in which customers are served. If we combine the customer potential with the business model steps, we can draw conclusions for the sales channel structure. Figure 94 shows the average recommendation for when selling should be undertaken personally vs. online, directly vs. indirectly or reactively.

Fig. 94 Sales channel structure per business model step

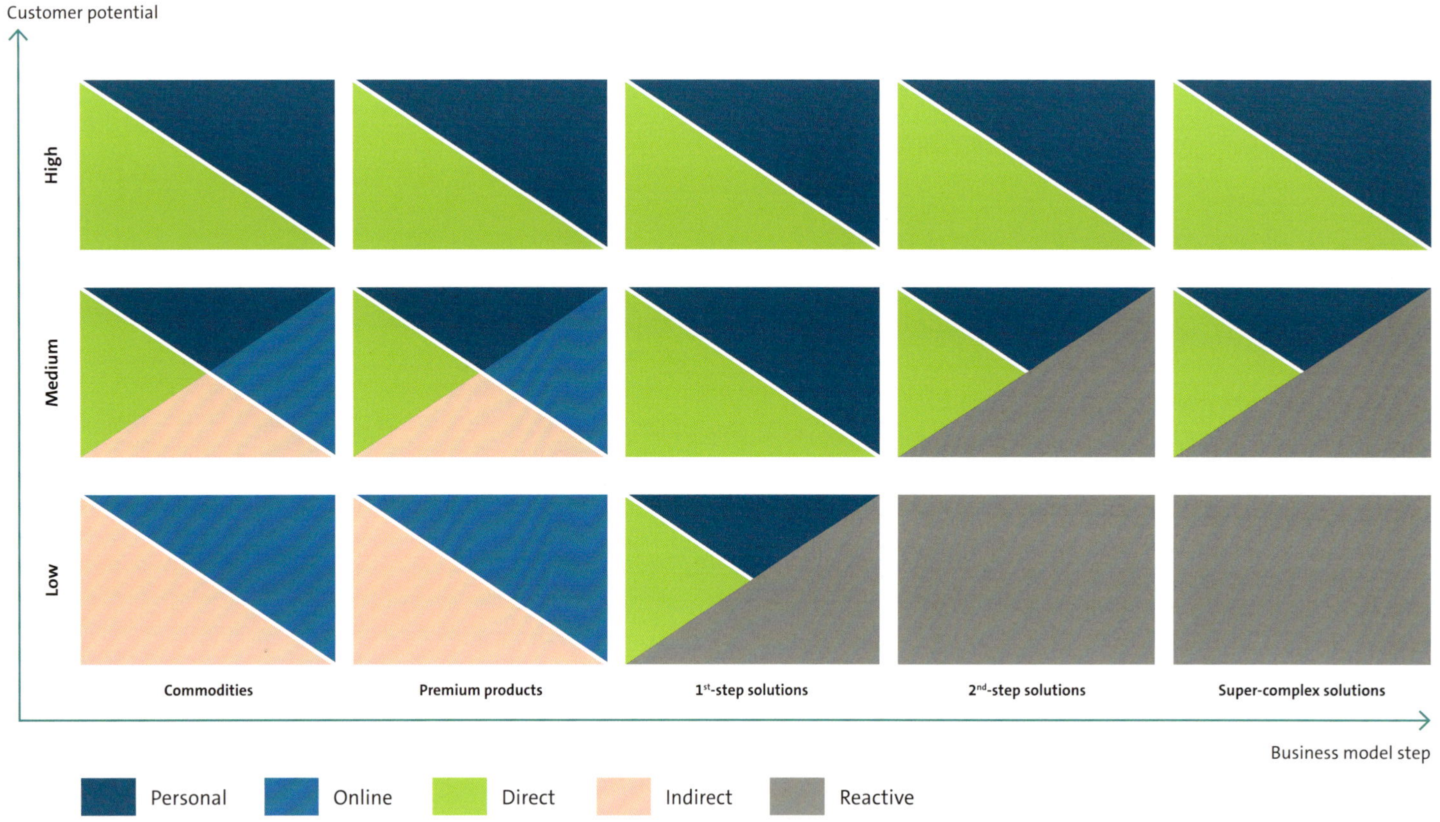

Let us go through this figure from top to bottom.

- **With high level of customer potential:**
  Direct, personal selling across all business model steps can prove worthwhile for high-potential customers. In the product business (commodity and premium), this contradicts the above recommendation of using online channels and selling indirectly. However, direct, personal selling is justifiable here if the potential that can be achieved clearly exceeds the costs to serve in selling. In this case, the higher expenses are worthwhile to increase the likelihood of acquisition. If online selling is chosen, it should be undertaken via one's own channels.

- **With medium level of customer potential:**
  A distinction should be made here. For example, both indirect and online sales channels may prove sensible for commodity and premium products. Direct, personal selling may also be worthwhile, depending on the case (see also deep dive on p. 317 regarding the question: *Is personal selling always fruitless in commodity markets?*).

  Due to the high costs, however, direct and personal selling are only worthwhile in specific cases for 2nd-step and super-complex solutions. Such a case would be if the respective customer reveals high-growth potential. Otherwise, a reactive sales approach is advisable here.

- **With low level of customer potential:**
  In this case, commodity and premium products should be sold with the greatest cost-efficiency possible, meaning primarily via (online) retailers.

  1st-step solutions are based on the integration of the customer's smaller process steps and require a certain degree of explanation. Since the services are relatively standardized and only adapted to a minor extent, however, comparatively low costs to serve are incurred. Direct, personal selling may therefore be worthwhile even in the case of a low level of potential if customers reveal sufficient growth potential. Otherwise, selling should be carried out reactively.

  2nd-step solutions and super-complex solutions necessitate extensive integration into the respective customer process. High costs to serve often arise for sales due to this integration and the necessary customization of the service. If the level of customer potential is low, these initial investments are unprofitable, meaning that selling should be reactive at best.

Step 2: Decisions regarding the establishment of a separate salesforce

In the second step, decisions regarding the establishment of a separate salesforce have to be made. The quick test above (p. 305) includes two crucial questions for this:

Question 1: Are the skills necessary for selling in the envisaged business model ***available*** in the existing salesforce?

It is a fallacy to believe that the best product salespersons are also outstanding solution salespersons. Many top product salespersons fail even when selling 1st-step solutions because the requirements are completely different. I dealt with the precise requirement profiles for sales employees in the solution business model steps in Chapter 9. Put bluntly, you need consultants who are familiar with the customer processes instead of salespersons specialized in products. Accordingly, sales employees have to come to grips with their customers much more thoroughly, ask the right questions, understand their processes and challenges, and be able to advise them accordingly. The supplier must therefore clarify whether its salesforce has these skills.

Question 2: Can the skills necessary for selling in the envisaged business model ***be learned*** by the existing salesforce?

If the necessary skills are not yet available but can be learned, the supplier must invest in the required training and coaching measures early on. We have often seen this happen too late and only half-heartedly. If you are convinced of the potential of your solution, you should also factor in the budget for necessary training and coaching measures.

If your answer to questions 1 and 2 is *no*, you should establish a separate, new salesforce to sell the solution. This also requires significant investment. You should also create the structures for establishing sales teams (see information box).

Information box: Decisions regarding team selling

The more complex the business model is structured, the more relevant the decisions regarding the sales teams become. This firstly involves the size of the sales team, which can range from lone wolves to a double-digit number of persons. Secondly, it involves the composition of the sales team.

Size of the sales team. As Figure 95 shows, the trend toward team selling continues to increase on the road to solution business models. Lone wolves are more likely to be found in low-complexity product selling.

Fig. 95 Trend toward team selling in the business model steps

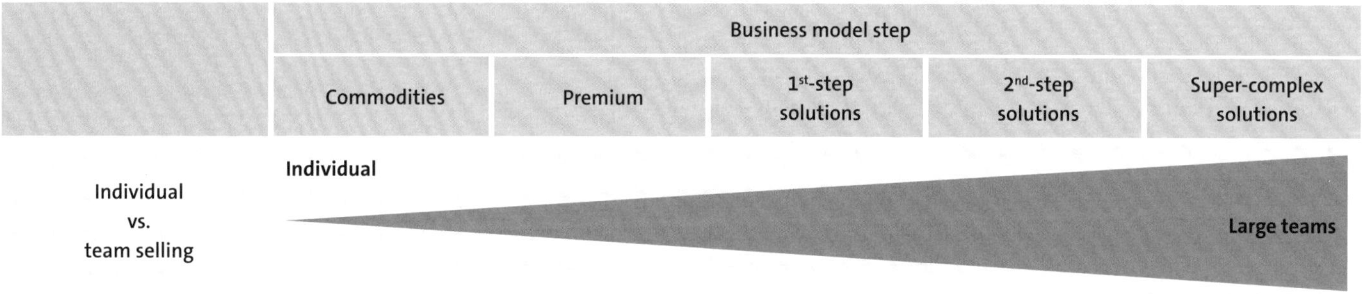

Composition of the sales teams. Not only the size of the sales teams, but also the number of roles needed in the team increase as the complexity of the solutions rises.

Let us again use the example of *Soarian*, which was developed as a super-complex solution for hospitals (see Chapter 6, p. 152). Which roles should be represented in this sales team? It is crucial that the roles are distributed so that the sales team as a whole has the following skills:

- Specific specialist knowledge. In this case, expertise in hospital procedures
- IT expertise
- Industry expertise. In this case, expertise regarding healthcare, ideally a background as a senior physician
- Executive-level standing, i.e., the seniority to be accepted on an equal footing by management board members and managing directors
- Commercial expertise
- Leadership skills for managing changing sales teams

Hardly any one person is able to offer all of these skills. The establishment of carefully composed sales teams is therefore critical, especially in the case of the 2nd-step and super-complex solutions. Added to this is the fact that multiple actors are also involved in the purchase decision on the customer's side. Serving them all exceeds the time capacities of any one person.

In order to deal with the buying center actors on an equal footing and with the same functional orientation, it is worthwhile mirroring them in the "selling center" (see Figure 96). This means that the composition of the selling teams is configured depending on the composition of the buying center. Since this has to be re-determined for each customer, the team structure has to be dynamic.

Fig. 96 Mirroring the buying center structures in the sales team

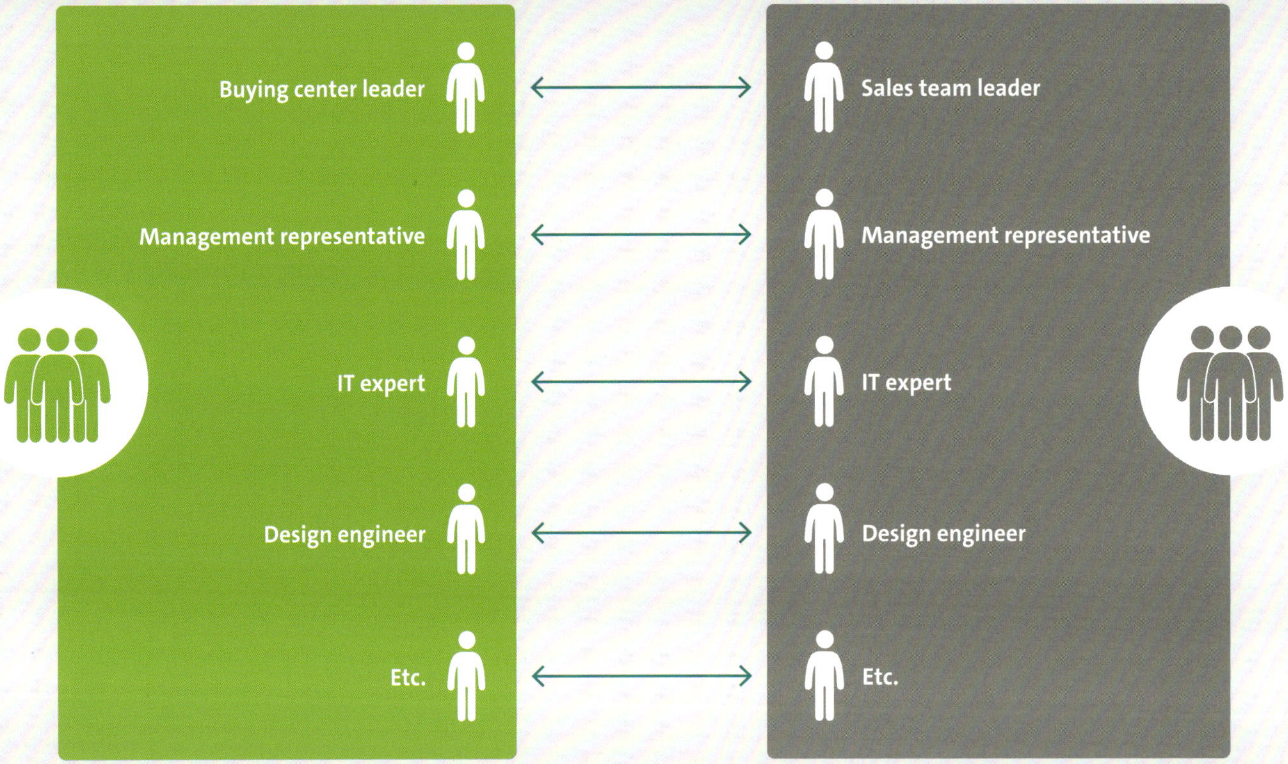

Step 3: Decisions regarding serving customers in specific segments

The third step involves structuring your sales based on the customer segments. Customer segments group existing and potential customers with specific purchasing needs (customers with demand for commodity products vs. customers with demand for individual solutions, for example). Customer segments can be used to serve individual customer groups more specifically. In our quick test for sales structures (p. 305), two questions are relevant:

Question 1: Is the business volume large enough to process individual customer segments with separate sales units?

If the business volume per customer segment is sufficiently large, segmentation can be carried out. If, for example, there are a sufficient number of customers with demand for individual solutions, it is worthwhile processing this segment with separate sales units.

Question 2: Is the necessary segment-specific know-how available in your salesforce?

The more complex the solutions that you sell become, the more important the salesforce's segment-specific know-how becomes. This relationship is made clear in Figure 97.

In more complex business models, sales employees have to be familiar with relevant industry standards, statutory regulations, and the value chains of their customers, and must take all of these into account when configuring the offerings. This particularly applies when you take over your customers' process steps with your solution. Detailed industry knowledge and speaking the same language also make communication with customers easier. It is therefore important to ensure that the existing salesforce has the necessary know-how concerning the corresponding customer segment.

If you answered *no* for question 2, recruiting sales employees with the segment-specific know-how is advisable. Or you can train suitable employees within the existing salesforce to become segment experts. If you are able to answer both questions *yes*, subdividing the salesforce according to relevant customer segments is advisable.

Fig. 97 The importance of segment-specific know-how

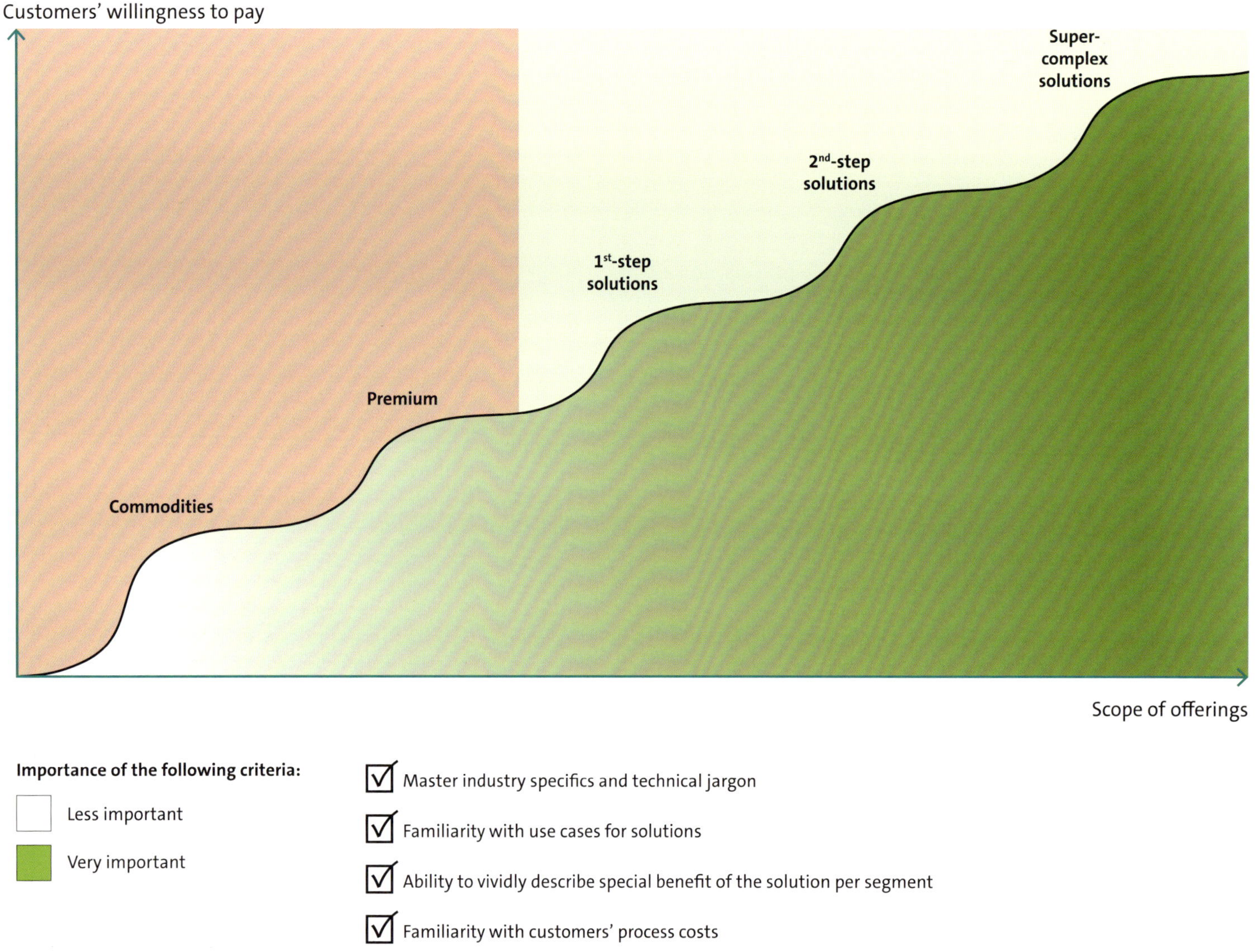

Deep dive: Is personal selling always fruitless in commodity markets?

If you proceed strictly according to the above assignments in Figure 93 (p. 308), the sales channel decision for commodity business models is simple and radical: A great deal can be processed online or via electronic data interchange. A lean dual-channel approach is best pursued by operating an inbound call center alongside the online channel, for instance. So are commodity salespersons a dying breed? According to some experts, they should have become extinct long ago. I recently heard the following statement at a management congress:

"It's like this for commodities: Certain minimum standards have to be met on the product side. If that's ensured, only the price then matters! That's why we no longer need personal selling here, but only a functioning online channel!"

A warning stop sign is appropriate at this point. Is this statement really correct?

Or is it too simple to believe that commodity salespersons will no longer exist in the near future? Is it conceivable that personal selling will remain interesting for at least part of this customer segment?

In that case, thinning out the commodity salesforce too quickly would be damaging. Dismantling a salesforce can be done quickly and comparatively easily. Establishing it again is anything but, however.

→ Will all commodity salespersons really die out immediately?

→ What happens during the transition time?

The answer is that the online channel should not be chosen for commodities in every case. The decision depends on the type of customer within the commodity segment.

With some types of customers, there is a transition time in which commodity salespersons are still of benefit. In order to make a more targeted sales channel decision, it is therefore worthwhile returning to the customer segment analysis.

Figure 98 shows the result of the customer analysis that we conducted for the commodity market of an energy module supplier. We analyzed the purchasing behavior and purchasing motives for 150 commodity customers. It revealed that a distinction can be made between four customer types according to two criteria: 1) "Convenience," or the desire to make little effort when purchasing, and 2) "price focus with commodities."

The figure shows how the customers of the company we analyzed are distributed over the customer types. The figure also shows the opportunity for personal selling for each type. In fact, the majority of the respective company's customers (40%) were assigned to type 1 and therefore have a strong online affinity. However, we still determined a good opportunity for personal selling with 20% of customers (type 2).

Fig. 98 Four customer types within the commodity segment (example: supplier of energy modules)

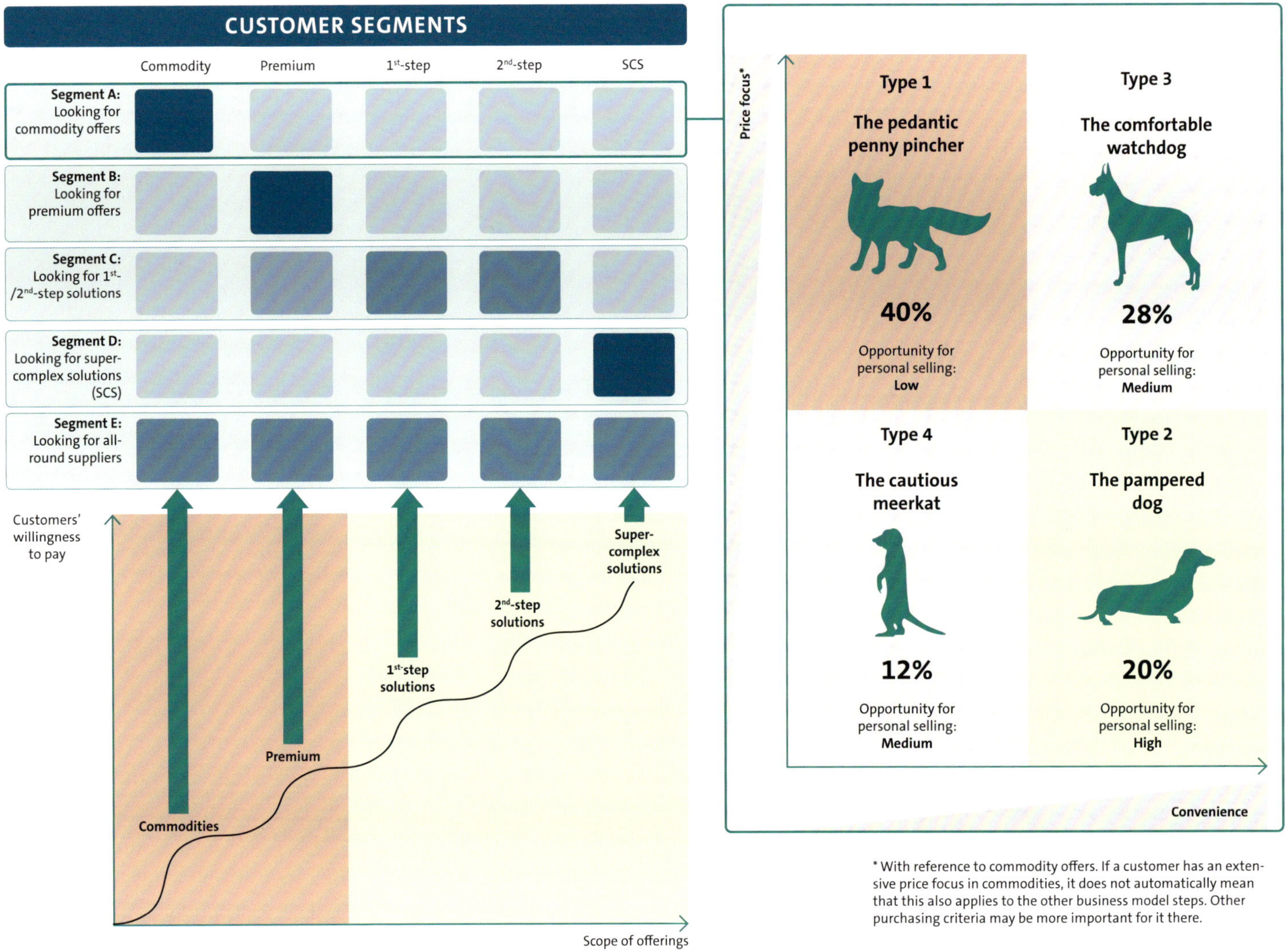

* With reference to commodity offers. If a customer has an extensive price focus in commodities, it does not automatically mean that this also applies to the other business model steps. Other purchasing criteria may be more important for it there.

The individual types are described in detail in Figure 99.

Fig. 99 Detailed description of the four customer types within the commodity segment

Price focus	Convenience: Low	Convenience: High
High	**Type 1 = the pedantic penny pincher** Price focus in the case of commodities: High Convenience: Low **Statement by which this type can be recognized:** *"We want the best price and invest time to find it, particularly in the case of commodities."* **Characteristics:** • Very cautious, works painstakingly and tactically when hunting for the best price • Professional price searcher, often with an internal purchasing department; perfectionistic • Stocks up when the price is low **Opportunity for personal selling:** Low **Recommendation:** Serve customers online.	**Type 3 = the comfortable watchdog** Price focus in the case of commodities: High Convenience: High **Statement by which this type can be recognized:** *"We have to be able to rely on a top price. We don't care how the supplier does it!"* **Characteristics:** • Works on a "transactional" basis with his suppliers; sets guidelines and is relaxed as long as these are kept to • Looks for suppliers that he can trust as regards the price, but switches immediately if this trust is broken • Permits personal selling if it reduces the transaction effort **Opportunity for personal selling:** Medium **Recommendation:** (1) Establish trust in your prices. (2) Ensure maximum convenience when purchasing.
Low	**Type 4 = the cautious meerkat** Price focus in the case of commodities: Low Convenience: Low **Statement by which this type can be recognized:** *"We don't search according to price, but according to reliability! We have to be able to rely on delivery deadlines, accessibility, guarantees, fast problem-solving, etc."* **Characteristics:** • Is concerned that his standards won't be met; seeks reliability • Attentively checks the situation; is not focused on the price but expects reliability in aspects such as fast availability and quality • Invests time in searching for reliability and guarantees **Opportunity for personal selling:** Medium **Recommendation:** (1) Earn trust as a reliable supplier. (2) Avoid salesperson fluctuation.	**Type 2 = the pampered dog** Price focus in the case of commodities: Low Convenience: High **Statement by which this type can be recognized:** *"Commodities such as C-parts mustn't cause us any effort. They shouldn't disrupt our core processes."* **Characteristics:** • Likes familiarity • Customer's business is doing well; there is no pressure to economize everywhere **Opportunity for personal selling:** High **Recommendation:** (1) Send salespersons to the customers to collect orders. (2) Check your sales costs. (3) Avoid salesperson fluctuation.

Let us take a closer look at the four types of customers in the commodity segment.

Type 1 = the pedantic penny pincher: The pedantic penny pincher is clearly the customer type with the greatest online affinity. For these customers, the price is central in the case of commodities. Convenience when purchasing is not a top priority for this customer type.

The pedantic penny pincher sometimes proceeds tactically when hunting for the best price. Accordingly, these customers frequently demonstrate professionalism while purchasing. Often, they have their own full-time purchaser or their own purchasing department. By nature, professional purchasers consistently strive to compare prices. When the price of a commodity product is low, they also stock up.

A supplier's sales employees have a tough job here. The opportunities for personal selling are few and far between. The above prognosis applies to the pedantic penny pincher: Classic online selling is widespread and will dominate universally in the future.

Type 2 = the pampered dog: This type is the complete opposite of type 1. This customer type wants to make little effort when purchasing. Price focus in the case of commodities is low.

The psychological effect via which a bond to this customer type arises is simple, but often underestimated: *Accustomation*. Many of these customers are accustomed to the fact that sales employees visit them in a certain rhythm and offer them commodities. This is the most convenient way to order something. At the same time, this accustomation represents a barrier to entry for other suppliers and other methods, and therefore not only for the competitors' sales employees, but also for new sales channels. Coming to grips with an online portal or software is a major obstacle for these customers.

A supplier of C-parts for car workshops described how it handles this type of customer as follows: *"There are workshops with shirt-sleeves processes. Here, we are serving the convenience and lack of structure in the customers' planning processes. They are accustomed to someone visiting them and replenishing their stocks."*

The recommendation for this type is therefore: Send salespersons to the customers to collect orders. There is one caveat, however: The customer revenues have to justify the sales costs.

Fluctuation in the supplier's field salesforce poses a major risk here. If the accustomation effect represents the psychological barrier for switching to the competition in this case, changes of sales employees are toxic. After all, the familiar sales employee is the face to which customers have become accustomed. We saw time and again in our analyses that the biggest risk of the customer's switching to the competition occurred shortly after a change of sales employee. This results, firstly, in the obvious recommendation of minimizing fluctuation in the salesforce. Secondly, a supplier can consistently monitor changes of competitors' salespersons and specifically address the customers concerned.

Type 3 = the comfortable watchdog: This type of customer scores high points in both criteria. They monitor prices precisely in the manner of a watchdog. At the same time, they like convenience in obtaining its commodity goods.

This type of customer works on a "transactional" basis with its suppliers. This means that they set guidelines for the cooperation. While supplier's services remain within these guidelines, the customer is relaxed and also open to new suggestions from the supplier. If the guidelines are violated, however, they switch suppliers without hesitation.

These customers' price focus and convenience lead them to occasionally check more closely. We have experienced companies that review all invitations to tender and prices down to the last detail once a year for individual C-part suppliers.

These customers remain loyal to the supplier if they trust its prices and purchasing is convenient. If the supplier reduces the transaction effort, the customer is open to personal selling.

Type 4 = the cautious meerkat: The price only plays a subordinate role for this type of customer. Nevertheless, they are prepared to make an effort when purchasing commodities. This effort is focused on reliability in terms of fast delivery and/or flawless quality, for instance.

One of the important characteristics of this type of customer is their concern that their standards will not be met. As a result, they are prepared to exert a great deal of effort in searching for reliability and guarantees.

Suppliers have to gain this customer type's trust as a reliable supplier. Once this trust has been established, personal selling also offers an opportunity for extensive customer loyalty.

The same recommendations as for type 2 apply as regards fluctuation in the supplier's salesforce. It is especially applicable in this case because type 4 customers form close bonds with individual sales employees. Fluctuation in the supplier's salesforce is toxic to these customers and opens the doors to the competition.

Step 4: Decisions regarding regional sales territories

Another fundamental decision that has to be made concerns regional territory subdivision in sales. In our quick test for sales structures (p. 306), we formulated two questions dealing with this. The rules are the same on all business model steps.

Question 1: Is the customer potential in the territories comparable?

Once again, the first measure for territory subdivision is the customer potential. It should be comparable in the sales territories. Employees with an excessively high customer density in their region are often overwhelmed by serving their customers, with the result that high-potential customers are neglected. Consequently, high sales volume potentials remain unexploited. In addition, selling new solution business models is often neglected in such territories. Employees with a low customer density in their region have too little work and are also demotivated. Territory subdivision can only function optimally with a balanced workload and comparable customer potentials.

Question 2: Is the territory breakdown optimized in terms of costs?

The second yardstick is the sales costs. Even the best subdivision cannot function in practice if the increases in sales volume are consumed by travel costs and costs to serve. It is worthwhile performing the *cost to serve analysis* to optimize the territory subdivision in cost terms. The supplier can use it to determine the cost to serve per customer and per sales territory. The estimated full costs serve as the basis. In this process, all activities with direct and indirect cost impacts that are required to serve a customer are totaled.

If at least one of the two questions is answered *no*, you should revise your territory subdivision again.

Step 5: Decisions regarding international sales territories

The fifth step deals with the question of whether entering the market in a new country is worthwhile for a business model. In our quick test for sales structures (p. 306), the following three questions are relevant in this regard.

Question 1: Is the market potential for the business model adequate in the target country?

In order to determine the market potential, a market analysis must be carried out for the relevant business model. If you already have your own salesforce on the ground for other business models, it can determine customer demand. If this is not the case, suppliers usually turn to external market research agencies.

Question 2: Are the barriers to entering the target country surmountable?

It is possible that there are serious barriers to entry in a country. These could be local customer preferences, political misgivings, or the general competitive situation. A detailed analysis, which I will go into shortly, is required for each business model in order to determine possible reservations early on and protect your company against bad investments.

Question 3: Does your salesforce have the necessary skills to sell the business model successfully on the ground?

Finally, you assess the internal skills available for implementation on site. The resource capacity, the logistical capabilities, and the intercultural skills of your own salesforce play a role in this.

If you have to answer question 1 or 2 *no*, entering the national market is not sensible. If you can answer both questions positively, but not question 3, cooperation with a sales company that is already working successfully for other companies in the location is appropriate. If you answer all of the questions affirmatively, entering the market with your own salesforce is sensible.

Analyzing barriers to entry. Often, success in new national markets is frustrated by barriers to entry. I would therefore like to introduce a detailed analysis for this in the following. It combines the SNOOFY factors of our business model steps (see Chapter 5) with the classic "PEST analysis" (see Figure 100).

The PEST analysis can be used to ascertain whether political/legal, economic, socio-cultural, or technological reservations exist in a target market.

Fig. 100 Barriers to entering the destination country – detailed analysis

	P Political/legal	E Economic	S Socio-cultural	T Technological
S				
N				
O				
O				
F				
Y				

Combining the individual dimensions of the PEST analysis with the SNOOFY factors results in the following four tables. They can be read as follows: The main political (or economic, socio-cultural, technological) topics are defined in the top line. The remaining lines contain questions that need to be asked before entering the target market. I would be happy to provide references and examples for the tables on request.

	Political/legal
Application to the SNOOFY factor	**Are there opportunities/risks for solution business models due to political/legal framework conditions?** **1.** Opportunities/risks due to the basic political system **2.** Opportunities/risks due to current political interference and current economic policy **3.** Opportunities/risks due to legal framework conditions **4.** Opportunities/risks due to tax law framework conditions
Servitization	**Re. 1:** *Are there risks in the long-term stability of the basic political systems that stand in the way of the high initial investments for some PASs and PDSs?* **Re. 2:** *Are there opportunities due to subsidy or investment programs for PASs and PDSs?* *Are there risks due to trade barriers for PASs and PDSs?* **Re. 3:** *Are there risks due to labor law restrictions for PPSs?* *Are there risks due to data protection law restrictions for PASs and PDSs?* **Re. 4:** *Are there risks due to tax law restrictions for certain PASs, PPSs, and PDSs?*
Nurture customer processes **O**perate processes for customers	**Re. 2:** *Are there risks due to trade barriers for nurture and operate?* **Re. 3:** *Are there risks due to laws that oppose the facilitation (nurture) or takeover (operate) of customer processes?* *Are there risks due to data protection law restrictions for nurture and operate?* *Are there risks due to technical trade barriers, e.g. as a consequence of deviating norms and standards?* *Are there risks in protecting intellectual property?* **Re. 4:** *Are there risks due to tax law restrictions for foreign suppliers and/or certain forms of nurture and operate?*
Outcome-based pricing	**Re. 1:** *Are there risks in the legal certainty of political institutions?* **Re. 2:** *Are there risks due to the introduction of statutory price brakes?* **Re. 3:** *Are there risks due to contract law restrictions for outcome-based pricing?* *Are there risks due to data protection law restrictions for outcome-based pricing?* **Re. 4:** *Are there opportunities/risks for alternative pricing models due to special accounting or tax law considerations?*
Far-reaching relationship	**Re. 1:** *Are there risks due to a lack of political protection against business practices such as corruption that oppose far-reaching cooperations between a supplier and customer?* *Are there risks due to restrictions caused by the inadequate political provision of market and customer knowledge, with the result that equal-status, fair cooperation is not possible?* **Re. 3:** *Are there risks due to the inadequate protection of intellectual property, which increases the risks of knowledge drain in the event of closer cooperation?* *Are there risks due to technical trade barriers, e.g. deviating norms and standards?*
Yield customer value-in-use	**Re. 4:** *Are there opportunities/risks for the development of the value proposition due to special accounting or tax law considerations?*

PPS – personally provided services (such as building cleaning) PAS – primarily automated services (such as automated production lines)

PDS – primarily data-driven services (such as software for device management)

	Economic
Application to the SNOOFY factor	**Are there opportunities/risks for solution business models due to economic framework conditions?** **1.** Opportunities/risks due to general macroeconomic framework conditions (economic cycle, exchange rates, wage costs, unemployment) **2.** Opportunities/risks as a consequence of the current market pressure in the target country: a. Market pressure in the direction of process efficiency due to low-price competitors b. Market pressure in the direction of unit sales growth due to high demand
Servitization	**Re. 1:** *Are there risks due to rising wage costs or low unemployment that necessitate investments in the development of PASs and PDSs?* **Re. 2a:** *Are there opportunities for PASs and PDSs or risks for PPSs as a result of high market pressure in the direction of process efficiency and effectiveness?* **Re. 2b:** *Are there opportunities for PASs and PDSs resulting from the industry's strategic focus on profitability (vs. growth)?*
Nurture customer processes **O**perate processes for customers	**Re. 1:** *Are there opportunities due to a general phase of economic growth, in which significantly more growth is enabled through nurture and operate?* **Re. 2a:** *Are there opportunities for reducing pronounced deficits in process efficiency and effectiveness with the aid of nurture and operate?* **Re. 2b:** *Are there opportunities for supporting the industry's strategic focus on growth/profitability with the aid of nurture and operate?*
Outcome-based pricing	**Re. 1:** *Are there opportunities due to the preference for higher liquidity/lower tied-up capital that can be achieved with the aid of leasing or pay-for-performance pricing models?*
Far-reaching relationship	**Re. 1:** *Are there opportunities to benefit from the economic upswing of other countries or safeguard against the weak economy in one's own country by means of specific partnerships?* **Re. 2a:** *Are there opportunities for reducing pronounced deficits in process efficiency and effectiveness with the know-how of partnerships?* **Re. 2b:** *Are there opportunities due to an investment backlog or industry development programs that can only be overcome with the aid of strategic partnerships and direct investments?* *Are there opportunities for cooperation in bidding consortia with the aid of potential strategic partners in the target country?*
Yield customer value-in-use	**Re. 1:** *Are there risks that long-term macroeconomic development could jeopardize long-term yield customer value-in-use?*

PPS – personally provided services (such as building cleaning) PAS – primarily automated services (such as automated production lines)
PDS – primarily data-driven services (such as software for device management)

	Socio-cultural
Application to the SNOOFY factor	**Are there opportunities/risks for solution business models due to socio-cultural framework conditions?** **1.** Opportunities/risks due to structural framework conditions: Educational system, population structure **2.** Opportunities/risks due to values, attitudes, or cultural reservations
Servitization	**Re. 1:** *Are there risks due to a lack of qualified personnel?* **Re. 2:** *Are there risks due to a perceived threat to jobs that lead to reservations toward PASs and PDSs?* *Are there opportunities/risks due to reservations toward or preferences for certain suppliers?* *Are there risks due to reservations toward the offered services?* *Are there risks due to reservations toward the origin of employees?*
Nurture customer processes **O**perate processes for customers	**Re. 1:** *Are there risks due to inadequately qualified personnel who have to be trained locally for nurture and operate?* **Re. 2:** *Are there risks due to socio-cultural attitudes and reservations toward nurture and operate?* *Are there risks due to reservations concerning necessary data provision, recording, and processing?* *Are there opportunities/risks as a result of prior positive/negative experiences with process support or takeover in the target country/target industry?*
Outcome-based pricing	**Re. 2:** *Are there risks due to socio-cultural attitudes and reservations toward alternative pricing models?* *Are there opportunities/risks due to higher risk propensity/ risk aversion in the target country?* *Are there risks due to the perception of increased lock-in and switching costs through contracts?* *Are there risks due to extremely pronounced skepticism?* *Are there opportunities/risks due to different perceptions of total prices?*
Far-reaching relationship	**Re. 1:** *Are there opportunities due to an investment backlog or industry development programs that have to be supported by trained specialists from abroad?* **Re. 2:** *Are there risks due to special cultural considerations concerning the establishment and maintenance of business relationships?* *Are there opportunities/risks due to the perception of long-term relationships as an economic added value?* *Are there risks due to pronounced ethnocentrism in the country?* *Are there risks due to extremely pronounced skepticism?* *Are there opportunities/risks due to general cultural misgivings regarding cooperations?*
Yield customer value-in-use	**Re. 2:** *Are there opportunities for high vertical integration due to socio-cultural preferences?* *Are there risks that prevent fulfillment of the implementation time up to the value proposition?* *Are there risks due to a lack of willingness to pay for intangible value propositions?*

PPS – personally provided services (such as building cleaning) PAS – primarily automated services (such as automated production lines)
PDS – primarily data-driven services (such as software for device management)

	Technological
Application to the SNOOFY factor	**Are there opportunities/risks for solution business models due to technological framework conditions?** **1.** Opportunities/risks due to the IT infrastructure **2.** Opportunities/risks due to the logistics infrastructure **3.** Opportunities/risks due to the mechanical infrastructure
Servitization	**Re. 1:** *Are there risks due to an inadequate IT infrastructure for PASs and PDSs?* **Re. 2:** *Are there risks due to an inadequate logistics infrastructure for PASs and PDSs?* **Re. 3:** *Are there risks due to an inadequate mechanical infrastructure for PASs and PDSs?* *Are there opportunities due to investment programs for technological development?*
Nurture customer processes **O**perate processes for customers	**Re. 1:** *Are there risks due to an inadequate IT infrastructure for nurture and operate?* **Re. 2:** *Are there risks due to an inadequate logistics infrastructure for nurture and operate?* **Re. 3:** *Are there risks due to an inadequate mechanical infrastructure for nurture and operate?*
Outcome-based pricing	**Re. 1:** *Are there opportunities for IT support for more complex contractual arrangements?* **Re. 3:** *Are there risks due to technical restrictions for outcome-based pricing?*
Far-reaching relationship	**Re. 1:** *Are there risks due to the spread of incompatible systems?* **Re. 3:** *Are there opportunities due to an investment backlog or major industry development programs that necessitate buying technology from abroad?*
Yield customer value-in-use	**Re. 3:** *Are there opportunities for using modern machines to generate a value proposition that is not offered in the target country?*

PPS – personally provided services (such as building cleaning) PAS – primarily automated services (such as automated production lines)
PDS – primarily data-driven services (such as software for device management)

SALES

WHY SALESTEGY?

STRATEGY – FINDING THE RIGHT BUSINESS MODELS

PRACTICAL EXAMPLES

SALES – IMPLEMENTING NEW BUSINESS MODELS IN SALES

Chapter 14
Adapting sales methods

Project steps

- Step 1: Using SPIN selling methods
- Step 2: Using Challenger Sale methods
- Step 3: Using customer inoculation in solution selling

Message of this chapter

The sales methods have to be reviewed for each new solution business model.

What should you have learned by the end of the chapter?

You are familiar with the sales methods of:

- SPIN selling
- Challenger Sale
- Customer inoculation in solution selling

DEATH VALLEY MARKETS

Strategy of minimum costs and prices

GREEN HILL MARKETS

Strategy of solution business models

This chapter deals with the sales methods that new business models necessitate. I explain three of these methods in greater detail (see Figure 101).

Fig. 101 Implementing new business models – project steps in sales management

Fig. 102 Classification of sales methods along the five business model steps

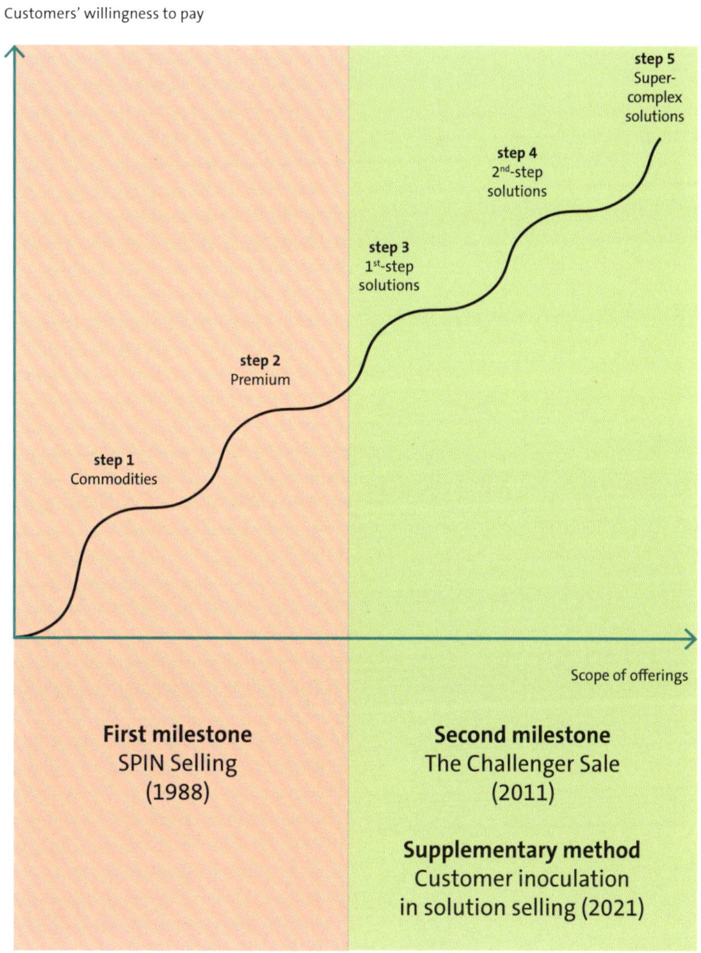

"Hard selling" approaches aimed at getting customers to quickly close a sale become ever less important as the scope of offerings increases. Conversely, sales methods that highlight the benefit and the added value for the customer are more appropriate for solution business models.

In practice, there are several terms for these sales methods, including "consultative selling," "value selling," and "insight selling." The sales landscape also offers training concepts that are used to improve solution selling. These include the *Miller-Heiman Method*, *Six Battlefields* from *MERCURI International*, and the *Clean Sales Method* from *Buhr & Team*.

Two of these sales methods are milestones. They pick up on value selling. Value selling is concerned less with selling the customer a product and/or a pure service and more with developing an added value for it. The milestones have been described in *SPIN Selling* by Neil Rackham and *The Challenger Sale* by Matthew Dixon and Brent Adamson. The difference lies in the application area for the business model steps (see Figure 102):

- SPIN selling originates from an age in which highlighting the benefit of premium products compared to commodities was important. Solution business models played a subordinate role.
- The Challenger Sale approach belongs to the epoch of solution business models. It focuses on the idea of delving deeply into the customers' business model and "challenging" it.

Project steps

Step 1: Using SPIN selling methods

The SPIN selling approach by Neil Rackham is a sales method that is based on systematic question techniques.[24] It particularly involves question techniques intended to encourage the customer to transform *implicit needs* into *explicit needs*. Implicit needs are unconscious problems expressed through vague dissatisfaction. As an example, Rackham cites slight dissatisfaction with the product in daily use or minor recurring problems (a machine that often stalls, for instance) that do not necessarily cause the customer to bring about specific changes. Conversely, *explicit needs* are strong motivations and the customer's willingness to immediately effect a change in order to solve the underlying problem. This requires the customer to be aware of this problem.

Rackham distinguishes between four phases of the sales pitch.

1st phase: Preliminaries

According to Rackham, the preliminaries phase is often overestimated in practice and does not play a crucial role in the likelihood of sales pitches being successful. Experience shows that failure in this initial phase is always preprogrammed if the sales employee outlines the product advantages at the beginning without being familiar with the customer's needs and problems. Rackham states that this phase should be as short as possible to use the time for the more effective, second phase of the customer dialog.

[24] The following content refers to the book *SPIN Selling* by Neil Rackham (1988), McGraw-Hill.

2nd phase: Investigation phase

According to Rackham, this is the most important phase because it primarily involves encouraging the customer to transform *implicit* needs into *explicit* needs. To be able to do this, the sales employee must have a precise understanding of the problems and their effects for the customer.

The SPIN selling approach is intended to help sales employees with four question forms:

- **S**ituation questions: These questions are aimed at obtaining a better and deeper understanding of the customer's actual situation: *"Which machines are you currently using for X production?" "At which points do you use them?"*
- **P**roblem questions: As the word indicates, this involves questions referring to potential problems in the customer's company: *"Do you have quality problems in the production process?" "How satisfied are you with the throughput time of machine X per minute?"*
- **I**mplication questions: These questions help in understanding the effects of the current problem for the company. The following question is possible here: *"What cost increases do machine downtimes cause per month?"* These questions are consciously problem-oriented and can lead to tense discussions. This is taken into account in the SPIN selling approach.
- **N**eed/payoff questions: The aim here is to show the added value of a solution by asking, for instance: *"How would the solution help you to increase the throughput time of the machine?" "How much would your result improve?"* The objective is to increase the customer's acceptance of the solution to the problem and to reduce objections. Unlike implication questions, need/payoff questions are solution-oriented.

3rd phase: Demonstrating capabilities

Both the advantages of the offered solution and the supplier's ability to help the customer implement the solution are shown in this phase of the sales pitch.

4th phase: Obtaining commitment

In the final phase of the SPIN selling approach, the sales employee must obtain a commitment from the customer. Apart from a clear purchase commitment or rejection, Rackham lists two further options here, one of which he classifies as a success and the other as a failure. The arrangement of a new appointment with the decision-maker is a success; the postponement of appointments to an uncertain point in time is a failure.

In our experience, sales employees particularly have difficulty with the **P**roblem and **I**mplication questions when learning the SPIN selling method. Particular attention should therefore be paid to how and at which points sales employees can use these questions in sales calls during training courses.

Step 2: Using Challenger Sale methods

Matthew Dixon and Brent Adamson, the authors of *The Challenger Sale*, emphasize the skills of the *challenger* salesperson type.[25] Above all, the Challenger Sale method has proved successful in complex selling situations. It can also be taught in training courses.

The challenger is characterized by an ability to communicate new findings to the customer. They provide findings that enable the customer to think about the company in a different way than before. They show the customer new ways of generating competitive advantages by identifying effectiveness and efficiency potentials. To do this, the challenger delves deeper into the customer's business model than the pure product salesperson. The challenger is also superior to the other salesperson types because they have the capability of assuming and maintaining leadership in the sales process.

According to Dixon and Adamson, challenger salespersons use what is called "commercial teaching" to communicate new findings. This approach consists of four fundamental rules:[26]

1st rule: Lead to your unique strengths. This rule involves the salesperson communicating the supplier's strengths to the customer. The supplier must know what it can do better than the competition.

2nd rule: Challenge customer's assumptions. In the Challenger Sale approach, the salesperson consciously questions the customer's assumptions. They communicate new concepts, perspectives, and findings to the customer. According to Dixon and Adamson, the majority of challenger salespersons understand the aspect of the customer business to which their offer refers better than the customer. *"A company that sells printers to hospitals, for example, may not know more about healthcare than the hospital administrators they sell to, but they certainly know more about information management in a hospital setting,"* claim the authors.[27] Rather than enthusiasm, reflectiveness is the best indication that the customer has taken new findings on board.

3rd rule: Catalyze action. Once the customer has begun to think things over, the challenger must motivate them to act. At this point, other salespersons would show the customer when purchasing the solution would be amortized. In commercial teaching, the customer is shown which profits are lost if the challenger's offer is *not* accepted.

4th rule: Scale across customers. One important element of commercial teaching is the selection of customers. Ideally, they are clustered so that a specific problem solution can be applied similarly by a high number of them. This leads to scalability for the salesperson, since they can quickly discover the customers' core problem during the sales calls.

According to Dixon and Adamson, good commercial teaching runs through six steps. The following figure shows the customer's stress level in the six phases. We can see that the customer moves through a period of low sentiment in the fourth phase and is gradually led to a positive stress level from this low point.

---

[25] The following content refers to the book *The Challenger Sale: How to Take Control of the Customer Conversation* by Matthew Dixon and Brent Adamson (2011), Portfolio / Penguin.

[26] See Dixon, Adamson (2011), pp. 55 ff.

[27] See Dixon, Adamson (2011), p. 60.

Fig. 103 The phases in the Challenger Sale approach

Phase	Description	Customer state
1. Warmer	Building credibility by reading their mind, demonstrating empathy	Intrigued
2. Reframe	First reframe of unrecognized problem, need, or assumption	Drowning
3. Rational drowning	Gradual intensification of the problem, both in degree and closeness to the customer	
4. Emotional impact	Psychological features of the problem, or presence in the individual's workflow, humanizing problem	Involved
5. Value proposition – a new way	A new framework for addressing the problem – implicitly tied to the supplier value proposition	
6. Your solution and implementation maps	Maps of supplier services or solutions linked back to the key teaching points; highlighted path to implementation	Relieved

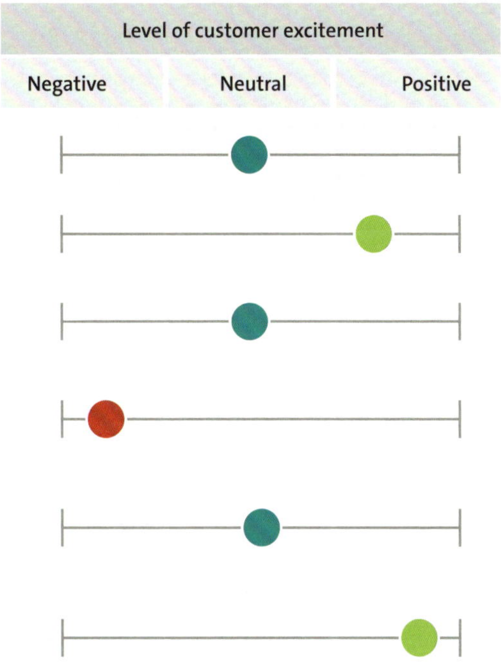

Source: Dixon, Adamson (2011), p. 66.

Step 3: Using customer inoculation in solution selling

"In the solution business, it is helpful to think through worst case scenarios beforehand in order to then achieve the best case scenario."
(Sales manager of the software supplier)

The more complex projects become, the greater the likelihood that they will not proceed without errors. It is worthwhile taking a look at the construction industry to illustrate this general rule: 98% of all complex, major construction projects veer out of control due to excess costs or delays. The average delay is 20 months.[28]

The general rule that complexity increases the likelihood of errors is not only true for the construction industry, it also applies to solution business models in all industries. This means that the more complex that solutions become, the greater the likelihood that they will not proceed without errors.

Let us remain with the construction sector for the sake of simplicity and put ourselves in the position of managers in customer companies who have to decide whether a solution offer is purchased or not. Just imagine the following scenario and consider what effect it would have on you:

You are the managing director of a medium-sized company that is currently planning to erect a new office building. Your company is now confronted by suppliers with a potpourri of options for products such as windows, facades, and furnishings. Added to these are various services that you are also being offered, including engineering and architectural services or the coordination of construction workers. If your company has no expertise in the construction sector, you will have difficulty assessing whether the suppliers' cost and time estimates will subsequently turn out to be realistic. Unfortunately, you have no choice but to trust the suppliers you have selected. In addition, you know that you will have virtually no chance of switching if they fail to deliver what you are expecting.

The uncertainty for you as the customer is tangible. Decision-makers are often confronted with such scenarios in complex solution business models. Particularly in the case of 2nd-step solutions and super-complex solutions, a high level of **customer uncertainty** arises due to this complexity and risks, which are difficult to assess. If we again pick up on the figure from the section on sales structures (see Figure 104), we are at the far right on the continuum.

[28] https://www.mckinsey.com/business-functions/operations/our-insights/the-construction-productivity-imperative; accessed on 11.09.2020.

Fig. 104 Approaches for reducing customer uncertainty

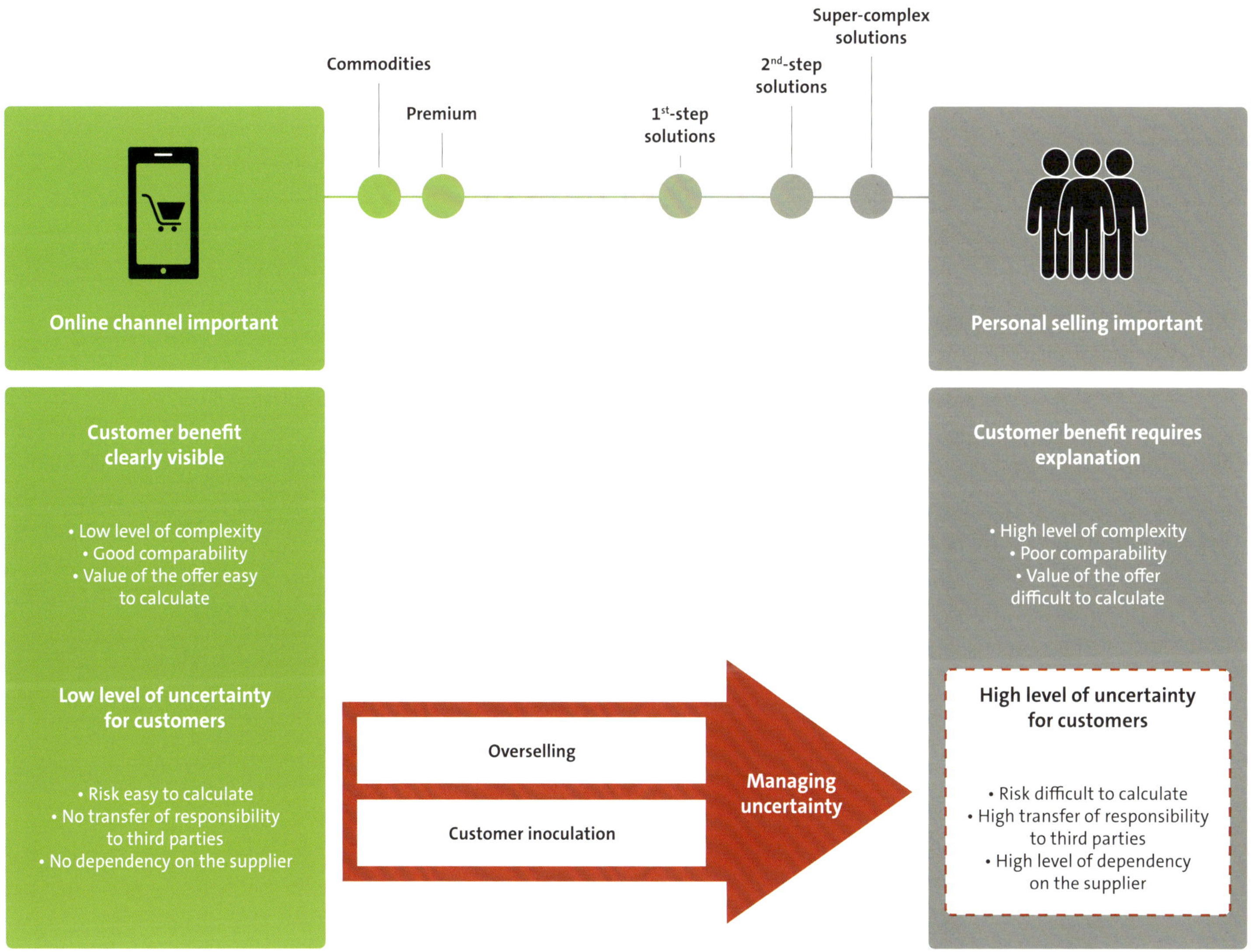

The question that we asked ourselves is: What can suppliers do in sales to reduce this customer uncertainty?

The sales method that is often used to attempt this can be referred to as "overselling" or "single-sided benefit selling." We have often observed this during our ride-alongs and coachings. Overselling means that the customer is given promises that are not kept. The managing director of a healthcare supplier defined overselling to me as follows: *"For me, overselling in the solution context begins when only the advantages and benefit elements are named, but possible difficulties and service errors are not disclosed at the same time, although the likelihood of their occurring isn't low. I understand that our customer advisors do it – they don't want to unsettle the customer and jeopardize contract completion. But I'm still against our doing it because we're lulling the customer into a false sense of security and therefore placing the customer relationship at risk in the long term."*

The more companies work with complex solution business models, the more the overselling approach proves to be a boomerang. This is due to the increased likelihood of errors in the solution business.

The managing director's statement makes three things clear:

1. The first sentence shows that the transitions to overselling are fluid.
2. The second sentence shows that overselling is often aimed at avoiding customer uncertainty.
3. The third sentence indicates the risk of a rude awakening.

So, if it is not worthwhile to conceal customer risks or detract from them, what can sales do to reduce customer uncertainty?

To achieve this, I would like to introduce the approach of customer inoculation in solution selling. The objective of customer inoculation is for the supplier to draw the customer's attention to possible disruptions prior to the provision of a solution service and to develop the best possible solution to the problem with it. This is carried out in five phases, which I briefly explain in the following.[29] A key question for sales can be assigned to each phase (see Figure 105).

[29] A more exhaustive description is available from: sales-management-department@ruhr-uni-bochum.de.

Fig. 105 The five phases of customer inoculation

Medical inoculation process

Phase 1:
Identification of potential pathogens

Phase 2:
Description of potential pathogens

Phase 3:
Selection of pathogens of high relevance

Phase 4:
Development of antibodies

Phase 5:
Selection of effective antibodies

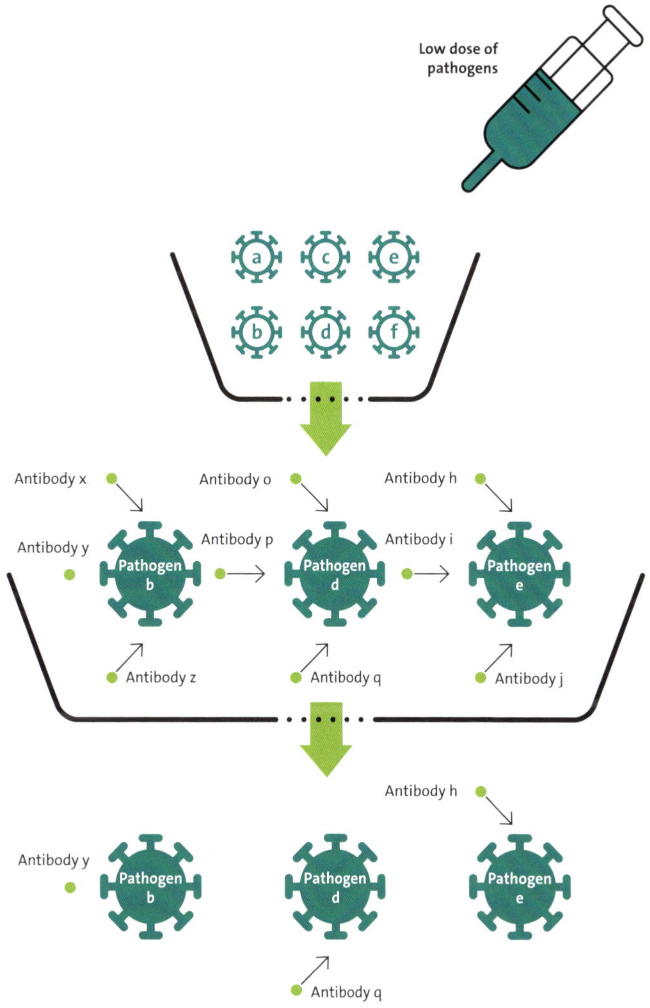

Customer inoculation in solution selling

Phase 1:
At which points of the solution process can complications arise?

Phase 2:
How can the complications in the solution process be described?

Phase 3:
Which of the solution problems could affect the customers' core processes most?

Phase 4:
What action options do the supplier and customer have to prepare for the possible complications?

Phase 5:
Which preventive measures should be prepared to be forearmed in the event of a complication?

1st phase: At which points of the solution process can complications arise?

The first phase initially involves going through and jointly anticipating the range of potential risks in the solution process. Each of the individual steps of the solution process is scrutinized. A suitable basis for this is a *solution blueprint*, or the visual representation of the process steps.

2nd phase: How can the complications in the solution process be described?

In this phase, the focus is placed on precisely describing conceivable problem situations and recording the individual elements of the respective problem from multiple standpoints. Both the customer and the supplier company must be clear about the effects a problem can have.

3rd phase: Which of the solution problems could affect the customers' core processes most?

This phase is concerned with naming the most serious disruptions that are most likely to occur during the course of the solution process.

4th phase: What action options do the supplier and customer have to prepare for the possible complications?

Once the problems have been clearly described and structured according to relevance, this step is concerned with showing as many ways as possible that contribute to solving the problem. It does not yet involve assessing the solution options, but it is an initial brainstorming session that produces as many action alternatives as possible.

5th phase: Which preventive measures should be prepared to be forearmed in the event of a complication?

The final step of the inoculation process involves determining the solution approaches for which a precise contingency plan should be developed in advance.

We have verified the effectiveness of customer inoculation in solution selling in several studies. The results have been published in a well-known research journal.[30] Our former PhD student Dr. Felix Brüggemann established the effectiveness of customer inoculation very vividly. The results of his doctoral thesis are shown in the following information box.

---

[30] Mikolon, Quaiser, Wieseke (2015), *Don't Try Harder: Using Customer Inoculation to Build Resistance against Service Failures*, Journal of the Academy of Marketing Science, 43 (4), 512–527.

Information box: Customer inoculation in 50,000 solution projects

On the basis of 50,000 solution projects, Dr. Brüggemann analyzed the extent to which sales employees pro-actively inoculated their customers in the sales process. This means the extent to which potential problems such as project delays and temporary downturns in performance were addressed and effective countermeasures (antibodies) were developed together with the customer.

In the evaluations, salespersons who used the concept of inoculation intensively were compared with salespersons who did not use it or hardly used it. The result was clear: The inoculation group's annual sales volume was around 17% higher than that of the others. The inoculation group's annual profits were also around 20% higher. In addition, the figures pertaining to pure solution sales volume (+7%) and allocated discounts (–6%) were also significantly better (see following table).

Improvements in performance thanks to customer inoculation			
+17% salesperson sales volume	+20% salesperson profit	+7% salesperson solution sales volume	–6% salesperson discounts

The following finding was also of central importance: Customer inoculation proves particularly useful in the case of highly complex offers such as solutions (see Figure 106). When an offer is less complex, customer inoculation has hardly any impact on the sales volume, whereas it leads to a 26% increase in sales volume in the case of highly complex offers.

Fig. 106 The effect of customer inoculation depending on the complexity of the offer

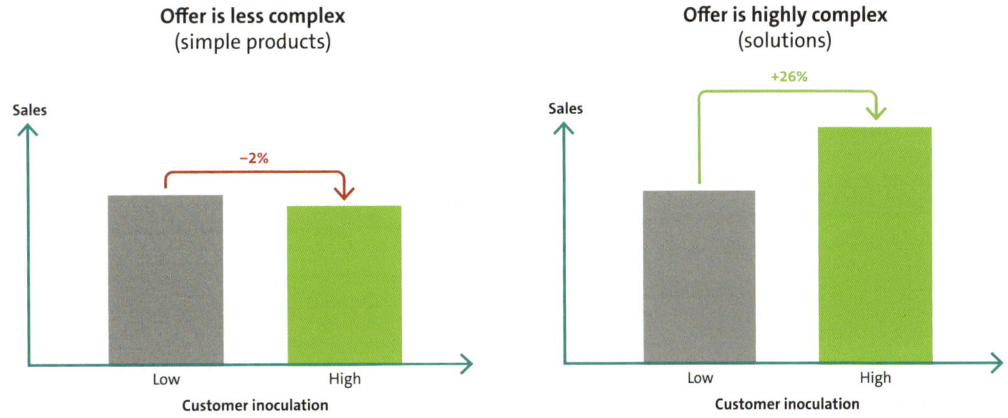

PREVIEW

Volume 1 (Jan Wieseke):

THE SALES PROFIT CHAIN

Understanding Causal Chains
Optimizing Sales
Increasing Profitability

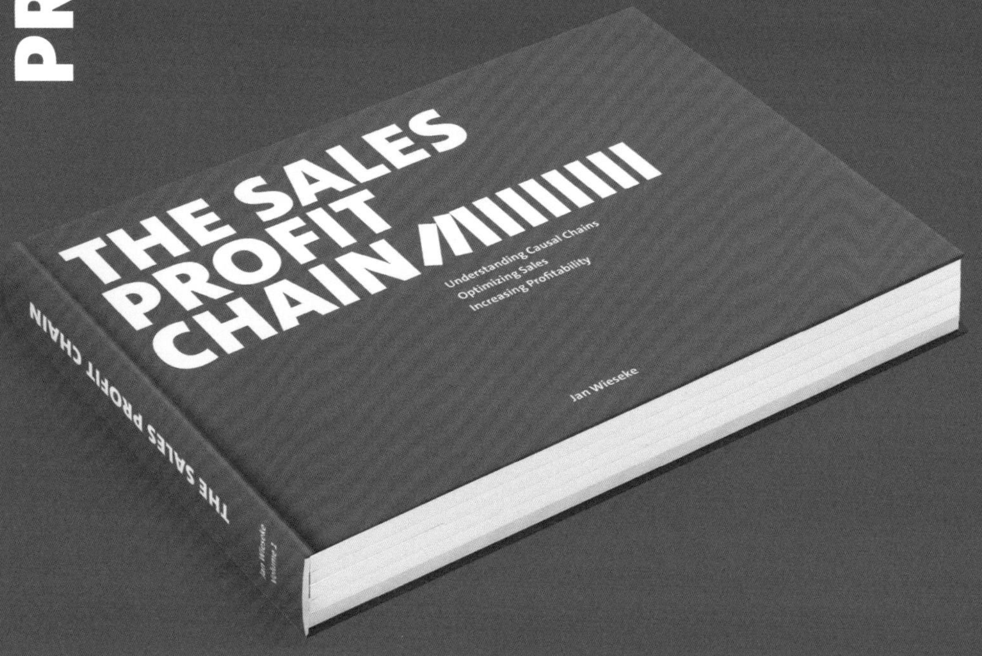

The Sales Profit Chain (SPC) has been developed for sales decision-makers. To accomplish this, we worked with more than 4,000 companies. The SPC approach uses 12 questions to analyze the most important causal chains of sales and therefore to access the most effective growth and efficiency levers.

WHAT IS NEW ABOUT THIS APPROACH?

- It uses a coherent overall sales logic (end-to-end approach).
- Instead of isolated, individual aspects of sales, the causal chains of sales management are made visible.
- The obstacles to implementing measures at the operational level are also explicitly overcome.

454 pages
ISBN 978-3-9823858-1-5
www.sales-profit-chain.com
sales-management-department@ruhr-uni-bochum.de